Springer Theses

Recognizing Outstanding Ph.D. Research

For further volumes:
http://www.springer.com/series/8790

Aims and Scope

The series "Springer Theses" brings together a selection of the very best Ph.D. theses from around the world and across the physical sciences. Nominated and endorsed by two recognized specialists, each published volume has been selected for its scientific excellence and the high impact of its contents for the pertinent field of research. For greater accessibility to non-specialists, the published versions include an extended introduction, as well as a foreword by the student's supervisor explaining the special relevance of the work for the field. As a whole, the series will provide a valuable resource both for newcomers to the research fields described, and for other scientists seeking detailed background information on special questions. Finally, it provides an accredited documentation of the valuable contributions made by today's younger generation of scientists.

Theses are accepted into the series by invited nomination only and must fulfill all of the following criteria

- They must be written in good English.
- The topic of should fall within the confines of Chemistry, Physics and related interdisciplinary fields such as Materials, Nanoscience, Chemical Engineering, Complex Systems and Biophysics.
- The work reported in the thesis must represent a significant scientific advance.
- If the thesis includes previously published material, permission to reproduce this must be gained from the respective copyright holder.
- They must have been examined and passed during the 12 months prior to nomination.
- Each thesis should include a foreword by the supervisor outlining the significance of its content.
- The theses should have a clearly defined structure including an introduction accessible to scientists not expert in that particular field.

Andreas W. Rost

Magnetothermal Properties near Quantum Criticality in the Itinerant Metamagnet $Sr_3Ru_2O_7$

Doctoral Thesis accepted by
University of St Andrews, UK

 Springer

Author
Dr. Andreas W. Rost
Scottish Universities Physics Alliance (SUPA),
School of Physics & Astronomy
University of St Andrews
North Haugh
St Andrews KY16 9SS, UK
ar35@st-andrews.ac.uk

Supervisor
Prof. Andrew Mackenzie
School of Physics & Astronomy
University of St Andrews
North Haugh
St Andrews KY16 9SS, UK
apm9@st-andrews.ac.uk

ISSN 2190-5053

e-ISSN 2190-5061

ISBN 978-3-642-14523-0

e-ISBN 978-3-642-14524-7

DOI 10.1007/978-3-642-14524-7

Springer Heidelberg Dordrecht London New York

Library of Congress Control Number: 2010931608

© Springer-Verlag Berlin Heidelberg 2010

This work is subject to copyright. All rights are reserved, whether the whole or part of the material is concerned, specifically the rights of translation, reprinting, reuse of illustrations, recitation, broadcasting, reproduction on microfilm or in any other way, and storage in data banks. Duplication of this publication or parts thereof is permitted only under the provisions of the German Copyright Law of September 9, 1965, in its current version, and permission for use must always be obtained from Springer. Violations are liable to prosecution under the German Copyright Law.

The use of general descriptive names, registered names, trademarks, etc. in this publication does not imply, even in the absence of a specific statement, that such names are exempt from the relevant protective laws and regulations and therefore free for general use.

Cover design: eStudio Calamar, Berlin/Figueres

Printed on acid-free paper

Springer is part of Springer Science+Business Media (www.springer.com)

Supervisor's Foreword

Our department nominated this thesis for a Springer award because we regard it as an outstanding piece of work, carried out with a remarkable level of independence. Andreas Rost joined us in 2005, as one of the inaugural Prize Students of the Scottish Universities Physics Alliance. Our research group has been working on $Sr_3Ru_2O_7$, in collaboration with our colleagues in the group of Professor Y. Maeno at Kyoto, since 1998. By early 2005 we had tantalising evidence that a novel phase was forming at very low temperatures, in an overall phase diagram dominated by quantum fluctuations.

We knew that comprehensive thermodynamic information would be needed in order to understand how this was happening, and that the demanding constraints of low temperature and high magnetic field meant that bespoke apparatus would need to be constructed. Andreas had studied the specific heat of glasses below 50 mK during his diploma thesis work at Heidelberg, and was brimming with ideas about how to proceed. We gave him advice, and constantly discussed the physics with him, but quickly realised that the best way to proceed practically was to give him a budget, and let him take the main design decisions, double-checking with us from time to time.

Andreas' design is notable for its flexibility. The same apparatus can be used to measure the specific heat by standard relaxation or a.c. methods, as well as the magnetocaloric effect in conditions ranging from quasi-adiabatic to highly non-adiabatic. All four modes were used during his measurements, in order to build up a comprehensive, cross-checked data set which allowed small entropy changes to be measured in absolute units.

The thesis reports three central pieces of new physics. First, the unusual 'liquid crystalline' behaviour of the correlated electrons in $Sr_3Ru_2O_7$ (established in separate work by our group) occurs in a bounded phase in an equilibrium phase diagram. Second, the entropic behaviour in the phase itself is unusual, suggesting the presence of some additional degrees of freedom whose origin is not yet understood. Third, the phase formation takes place against a background of strongly peaking entropy, suggesting that quantum criticality plays a key role in the physics. Some of these key results were published in 2009 as a full-length

Research Article in Science magazine, and several more manuscripts are in preparation. Already, however, the work has led to surprising and intriguing connections beyond our immediate field. The peaking of the entropy is surprisingly strong, and raises the question of whether the phase forms to avoid an entropic singularity. Such a singularity is not predicted by traditional theories of quantum criticality, but is, apparently, a feature of novel approaches in which string theory techniques are applied to the problem. Time will tell if this is a promising scientific train of thought, but Andreas' data being featured in conferences on cosmology is not something that any of us could have predicted when this project began. It illustrates the fact that the rewards for thorough, high precision experimental work are many and varied.

St Andrews, April 2010   A. P. Mackenzie and S. A. Grigera

Acknowledgments

The work presented in this thesis would not have been possible without the help and support from a number of people. First and foremost I am greatly indebted to my co-supervisors Andy Mackenzie and Santiago Grigera, who introduced me to the fascinating field of correlated electron physics and always had an open door for my plentiful questions. Not only did they support me in all aspects of the work but I am particularly grateful for the many opportunities I was given to explore science beyond the necessities of the project. Special thanks are due to Rodolfo Borzi, Naoki Kikugawa and Robin Perry, who helped by introducing me to the experimental equipment in the laboratory and the measurement techniques used by the group. Their patience with me and their help with the day-to-day work in the early parts of my Ph.D. was invaluable. Discussions with them helped me understand many of the small but important details of the scientific research undertaken by the laboratory. I need to particularly thank Robin Perry together with Jean-François Mercure and Alexandra Gibbs for the sample growth and characterisation. The work presented here would not have been possible without this. I shared all the little successes and failures occurring during a Ph.D. with Jean-François Mercure, Andrew Berridge and Jason Farrell, who carried out their Ph.D. studies at the same time as me. Most of the results presented here are crucially informed by long discussions with them and I learned a lot from their insights into $Sr_3Ru_2O_7$ and the work carried out by them. I also benefited from interesting discussions with many other members of the lab and I want to thank Jan Bruin, Demian Slobinsky, Alexandra Gibbs as well as the members of other groups—Ed Yelland, Alex Holmes, Marie-Aude Measson, Felix Baumberger and Anna Tamai—for their time and support. One aspect that made St Andrews a special place for a Ph.D. project was the interaction with other groups in the department. Many hours of discussions on the technical, theoretical and sometimes philosophical aspects of my work were spent in the common room. I am particularly grateful to Chris Hooley, Andrew Green and John Allen for sharing their insights into physics with me. I would also like to use the opportunity to thank the technical staff in the department. Reg Gavine not only ensured the constant supply of liquid helium but also helped with many other aspects of laboratory life, such as failing pumps.

Further thanks are due to Mark Ross in Electronics as well as David Steven and George Robb from the workshop for putting my designs into reality. I especially would like to thank my parents as well as my sister and brother-in-law, who always support me in my studies and often help me to see what is important beyond the work in the lab. Finally and most importantly I would like to thank Alex for accompanying me along the journey of my Ph.D. time. Her constant support, encouragement and understanding are invaluable.

The work reported in this thesis was financially support by the Scottish Universities Physics Alliance (SUPA) Graduate School and the Engineering and Physical Sciences Research Council (EPSRC). I also benefited from the travel grant scheme of the Scottish International Education Trust (SIET), allowing me to visit the laboratories of Y. Maeno at Kyoto University.

Contents

1	**Introduction**		1
	References		4
2	**Background Physics**		7
	2.1	Itinerant Electron Systems	7
		2.1.1 Non-interacting Electron Theory	8
		2.1.2 Magnetic Field Effects	12
		2.1.3 Electric Transport	17
		2.1.4 The Fermi Liquid	18
		2.1.5 Beyond the Fermi Liquid	20
	2.2	The Physics of the Ruthenate Family	24
		2.2.1 Crystal Structure and Synthesis	24
		2.2.2 Thermodynamic Properties and Magnetic Phase Diagram	25
		2.2.3 Electronic Structure Properties	34
	2.3	Summary	41
	References		42
3	**Thermodynamic Measurements of Entropy**		45
	3.1	General Considerations on Thermodynamics in Magnetic Fields	46
		3.1.1 The Laws of Thermodynamics for Magnetic Systems	46
		3.1.2 Phase Transitions	49
	3.2	Experimental Consequences	50
		3.2.1 Principle of Specific Heat Measurements	51
		3.2.2 Magnetocaloric Measurements	56
	References		63
4	**Design and Characterisation of Novel Experimental Setup**		65
	4.1	Measurement Environment and Sample Holder	65

	4.2	Design of Experimental Setup..................................	67
		4.2.1 Design Goals and Experimental Realisation............	67
		4.2.2 Sample Platform and Thermal Bath....................	68
		4.2.3 Thermometer..	70
		4.2.4 Heater...	71
		4.2.5 Estimates of Thermal Performance...................	72
	4.3	Thermometry...	75
		4.3.1 Physical Properties of Thermometers................	75
		4.3.2 Thermometer Calibration..............................	78
	4.4	Characterisation Run With Sr_2RuO_4........................	82
		4.4.1 Sample..	83
		4.4.2 Specific Heat in Field..................................	83
		4.4.3 Specific Heat at Zero Field............................	86
		4.4.4 Magnetocaloric Oscillations...........................	86
	4.5	Characteristics and Details of Measurements on $Sr_3Ru_2O_7$.....	88
		4.5.1 Sample Selection.......................................	88
		4.5.2 Thermal Link...	89
		4.5.3 Specific Heat...	90
	References..		91
5	**Experimental Results and Discussion**............................		93
	5.1	Caloric Studies of Magnetic Phase Transitions in $Sr_3Ru_2O_7$....	94
		5.1.1 Evolution of Entropy across Phase Transitions as a Function of Field....................................	94
		5.1.2 Specific Heat Signature of Phase Transitions.........	101
		5.1.3 Discussion..	104
	5.2	The Low and High Field States of $Sr_3Ru_2O_7$.................	112
		5.2.1 The Low Field Fermi Liquid State...................	112
		5.2.2 The High Field Fermi Liquid State..................	117
		5.2.3 Discussion..	119
	References..		129
6	**Conclusions and Future Work**.....................................		133
	References..		137
7	**Appendices**..		139
	7.1	Appendix A: Material Properties.............................	139
	7.2	Appendix B: Angular Dependence of the Magnetocaloric Signal..	139
		7.2.1 Study at 19°...	141
		7.2.2 Study in the *ab*-Plane...............................	142
	References..		144

Chapter 1
Introduction

In solid state physics one is in the situation that it is possible to write down the Hamiltonian governing the physics of the systems under investigation. The quantum mechanical Schrödinger equation describing the electrons and nuclei that form the solid contains their kinetic energy terms and the Coulomb interactions between the particles. This situation means effectively that one has a 'theory of everything' available, describing in principle all material properties below the energy scales of solidification. From a reductionist point of view this is a *fait accompli*.

Unfortunately, for the number of particles in realistic solid state systems this 'theory of everything' is a highly complex set of equations that currently is not only insoluble analytically but also beyond the capability of numerical methods in most cases.[1] It is in particular the Coulomb interaction between particles that makes the equations intractable and correlates the motion of all electrons with one another. One of the most radical simplifications of the equation is simply to neglect the Coulomb interaction, leading to the electrons conforming to the theory of the Fermi gas. Surprisingly it is found that the functional form of the properties of a wide range of crystalline materials can be successfully described by this theory. A reason for this was provided by Landau who showed that under certain general conditions the fundamental low energy excitations of an interacting Fermi liquid have the same charge and spin quantum numbers as non-interacting electrons, explaining the similar behaviour to the Fermi gas. However the extraordinary

---

[1] One has to point out though that developments in numerical calculations have made great advances in recent years and are now a useful tool in the understanding of band structures of crystalline materials. Especially the availability of tool sets such as Wien2K [1] and CASTEP [2] make the use of the techniques available to non-experts and become more and more used in the lab environment. However, they are based on certain approximations of the full Hamiltonian and therefore currently unable to treat problems going beyond these approximations.

A. W. Rost, *Magnetothermal Properties near Quantum Criticality in the Itinerant Metamagnet $Sr_3Ru_2O_7$*, Springer Theses, DOI: 10.1007/978-3-642-14524-7_1,
© Springer-Verlag Berlin Heidelberg 2010

success of Fermi liquid theory makes materials not falling into this category of particular scientific interest.

If the effective interactions between the quasiparticles in a Fermi liquid become strong enough, its basic assumptions can break down and this opens the path to a wealth of unexpected phenomena and the continuing discoveries of new physics. This makes solid state physics an extraordinary field where experimental discoveries lead to effective theoretical descriptions, as happened in the case of superconductivity and BCS theory, and at the same time theoretical developments are guiding experiments, with a recent example being the search for p-wave superconductivity in Sr_2RuO_4 [3]. But the wealth of phenomena extends far beyond unconventional superconductivity to a diverse range of physics such as is observed in Kondo systems [4, 5], the fractional quantum Hall effect [6, 7] or the rapidly growing field of frustrated spin systems [8].

One route to novel physics in metallic systems that has been investigated intensely in recent years is the phenomenon of quantum criticality—the physics related to a zero temperature second order quantum phase transition that can be crossed by tuning an external parameter such as pressure or magnetic field. In the vicinity of this phase transition—called the quantum critical point—the thermodynamic properties are not dominated by classical thermal fluctuations, as are finite temperature second order phase transitions, but by quantum fluctuations. These renormalise the properties of the two competing quantum ground states, potentially giving rise to non-analytic singularities of their thermodynamic properties at the quantum critical point. This allows for novel quantum phases, that are under normal conditions thermodynamically unstable, to emerge in the vicinity of quantum critical points.

An experimentally successful approach to quantum critical points in phase diagrams has been the suppression of a finite temperature second order phase transition to zero temperature [9] by tuning such variables as pressure [10], magnetic field [11] or doping [12]. Figure 1.1 shows the example of $CePd_2Si_2$ studied by Mathur et al. [10]. Here the observed phase transitions in the temperature–pressure phase space are shown. At zero pressure the material undergoes an antiferromagnetic second order phase transition as a function of temperature. If one applies pressure the antiferromagnetic transition temperature T_N is systematically suppressed towards lower temperature. However the quantum critical point that would ultimately be reached at a critical pressure p_C is 'avoided' by the material becoming superconducting inside the indicated dome. What makes this new superconducting phase so interesting is that it is supposed to be mediated by electron–magnon rather than electron–phonon interactions. Quantum criticality is also thought to play a significant role in the overall phase diagram of a range of systems of which two of the most well known are the cuprate high-T_C superconductors and f-electrons in heavy Fermion Kondo systems [9].

The material studied in this project, $Sr_3Ru_2O_7$, is an example of a metallic system thought to be close to a quantum critical phase transition that can be tuned by magnetic field. It has a currently unique combination of properties that make it a particularly interesting example.

1 Introduction

Fig. 1.1 Here the phase diagram of $CePd_2Si_2$ as a function of temperature and pressure is reproduced from Ref. [10]. The transition temperature T_N of the second order phase transition to an antiferromagnetic phase is suppressed towards lower temperatures by applying pressure. Before the quantum critical point is reached the material undergoes a further transition to a state of unconventional superconductivity thought to be mediated by magnons rather than phonons. The transition temperature T_C to superconductivity has been multiplied by a factor 3 for the purpose of presentation

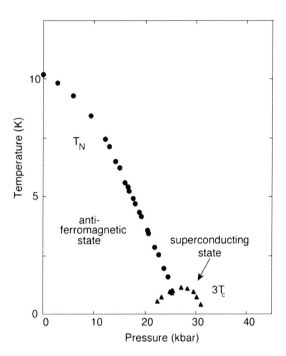

First, quantum criticality in $Sr_3Ru_2O_7$ is approached by suppressing the critical end-point of a line of first order phase transitions to zero temperature leading to a novel class of quantum phase transition—the quantum critical end-point. The tuning parameters here are the magnitude of the magnetic field and its orientation with respect to the crystal structure, with the quantum critical end-point being realised with the magnetic field being applied along the crystallographic c-axis.

Secondly, the associated physics can be studied in clean single crystal samples whose electronic structure is quasi-two-dimensional. This allows for a detailed study of the fermiology of the quantum mechanical ground states in the vicinity of the proposed quantum critical point by a range of otherwise not usable experimental techniques such as quantum oscillations, angular resolved photo emission studies (ARPES) or scanning tunneling microscopy (STM).

Last but not least it was found in ultra clean samples with mean free paths of the order of 3000 Å that in the vicinity of the expected quantum critical end-point the material forms a novel quantum phase whose electronic system shows nematic-like transport properties.

This combination of experimental findings makes $Sr_3Ru_2O_7$ an ideal candidate to study the above phenomena in detail. Though a wealth of experimental studies have been performed, so far no direct measurement of the evolution of the entropy of the system across the low temperature phase diagram has been achieved. This is a major gap in our knowledge since the entropy encodes important information on the overall degrees of freedom of the system.

The measurements presented in this thesis have two main purposes. First, the phase diagram of the novel quantum phase is investigated. Here specific heat measurements are used to identify the thermodynamic nature of the surrounding phase transitions as well as to study the temperature dependence of the entropy inside the phase. The second and equally important focus of the project is the change of entropy of the surrounding 'normal' phases upon approaching the critical region as a function of magnetic field. The method normally employed to reconstruct the entropy at a given magnetic field H and temperature T is to integrate the specific heat at H from zero temperature to T. In the case of approaching a quantum critical point this cannot be done since finite temperature specific heat measurements cannot reliably be extrapolated to zero temperature. Therefore, here the less commonly encountered magnetothermal effect was used. In this measurement the entropy of a system is extracted by analysing the temperature changes of a sample during a magnetic field sweep.

In the first part of the following chapter the main assumptions and results for the theory of the Fermi gas and the Fermi liquid are reviewed insofar as they are relevant to this project. This is followed by a brief introduction to quantum criticality and its importance to the study of materials that go beyond the Fermi liquid. The second part will review the physical properties of $Sr_3Ru_2O_7$ and introduce the current experimental evidence for quantum criticality, the existence of a novel quantum phase and the electronic structure in the surrounding 'normal' states. Chapter 3 introduces the relevant aspects of thermodynamics in magnetic fields and the measurement methods for specific heat and the magnetothermal effect. In particular the strong magnetic moment and the requirements for the magnetothermal measurement made it necessary to design and characterise a new experimental setup and calibrate thermometers over the relevant temperature and magnetic field range. In Chap. 4, the details of this experimental apparatus together with characterisation measurements are presented. This is followed in Chap. 5 by the results of the magnetothermal study of the entropy evolution as a function of temperature and magnetic field in the vicinity of the critical region of the phase diagram. In particular, results on the anomalous new phase and the evolution of the normal state upon approaching the critical field will be presented and discussed in relation to previous measurements. Furthermore, the results of a preliminary angular study with the magnetic field being applied at directions different from the crystallographic c-axis are given in Appendix B. The thesis will conclude with a brief summary of the achievements of the project.

References

1. Blaha P, Schwarz K, Sorantin P, Trickey SB (1990) Full-potential, linearized augmented plane-wave programs for crystalline systems. Comp Phys Commun 59(2):399–415
2. Segall MD, Lindan PJD, Probert MJ, Pickard CJ, Hasnip PJ, Clark SJ, Payne MC (2002) First-principles simulation: ideas, illustrations and the CASTEP code. J Phys Condens Matter 14(11):2717–2744

References

3. Mackenzie AP, Maeno Y (2003) The superconductivity of Sr_2RuO_4 and the physics of spin-triplet pairing. Rev Mod Phys 75(2):657–712
4. Stewart GR (2001) Non-Fermi-liquid behavior in d- and f-electron metals. Rev Mod Phys 73(4):797–855
5. Kondo J (1964) Resistance minimum in dilute magnetic alloys. Prog Theor Phys 32:37
6. Tsui DC, Störmer HL, Gossard AC (1982) Two-dimensional magnetotransport in the extreme quantum limit. Phys Rev Lett 48(22):1559–1562
7. Laughlin RB (1983) Anomalous quantum Hall-effect—an incompressible quantum fluid with fractionally charged excitations. Phys Rev Lett 50(18):1395–1398
8. Diep HP (2005) Frustrated spin systems. World Scientific Publishing, Singapore, London
9. Mathur ND, Grosche FM, Julian SR, Walker IR, Freye DM, Haselwimmer RKW, Lonzarich GG (1998) Magnetically mediated superconductivity in heavy Fermion compounds. Nature 394(6688):39–43
10. Löhneysen Hv, Rosch A, Vojta M, Wölfle P (2007) Fermi-liquid instabilities at magnetic quantum phase transitions. Rev Mod Phys 79(3):1015–1075
11. Custers J, Gegenwart P, Wilhelm H, Neumaier K, Tokiwa Y, Trovarelli O, Geibel C, Steglich F, Pepin C, Coleman P (2003) The break-up of heavy electrons at a quantum critical point. Nature 424(6948):524–527
12. Schröder A, Aeppli G, Coldea R, Adams M, Stockert O, von Löhneysen H, Bucher E, Ramazashvili R, Coleman P (2000) Onset of antiferromagnetism in heavy-Fermion metals. Nature 407(6802):351–355

Chapter 2
Background Physics

In this chapter I will introduce the theoretical concepts and experimental data that are relevant to the study of $Sr_3Ru_2O_7$ presented in this thesis. From a theoretical point of view the fundamental phenomena of a wide class of correlated electron systems are surprisingly well described by a non-interacting 'Fermi gas' theory. One of the first assumptions of this theory is not to take into account any electron–electron interactions besides Pauli's Exclusion Principle. It is thanks to a theory by Landau [1–3] that we understand why the results for the Fermi gas also apply qualitatively to strongly interacting Fermi liquids. Since the properties of the metallic ground state of $Sr_3Ru_2O_7$ in zero magnetic field are well described by the concepts of Landau's Fermi liquid theory I will review its most important results in the first part of this chapter. Indeed the predictions of Landau's Fermi liquid theory are so robust that it is in particular the systems that go beyond that theory and show non-Fermi liquid behaviour that have attracted a wealth of experimental and theoretical, with the high temperature superconductors of the cuprate [4] and recently iron–pnictide families [5] being but two of the most famous examples. In particular, the concept of quantum criticality has been used experimentally as a route to novel quantum states and will be briefly discussed.

The second part of this chapter will review the most relevant experimental data existing to date on $Sr_3Ru_2O_7$. Here I will in particular discuss the thermodynamic evidence for a metamagnetic quantum critical end point as well as the existence of a novel quantum phase in its vicinity. The chapter will finish with a summary of the experimental evidence for Fermi liquid quasiparticle excitations across the phase diagram.

2.1 Itinerant Electron Systems

To give a thorough introduction to the 'Fermi gas' or Landau's Fermi liquid theory is beyond the scope of this thesis. However, in order to put the experimental results and their significance into perspective it is necessary to summarise the most relevant results of the non-interacting theory in the first part of this section. Here particular emphasis will be put on the properties in magnetic fields. This will be followed by a basic introduction to Fermi liquid theory in order to show which concepts form the basis for its successful description of a wide range of itinerant electron systems. For more detailed discussions and derivations of the results presented here the books by Ashcroft and Mermin [6] for the properties of solid state systems in general and by Abrikosov et al. [7] for Landau Fermi liquid theory in particular give a good treatment of the subject and provide further references to the literature.

2.1.1 Non-Interacting Electron Theory

Many metals share a range of similar properties such as a linear temperature dependence of the specific heat at low temperatures. A very important step in understanding the generality of the physical laws applying to a vast range of different materials was Sommerfeld's theory of non-interacting fermionic particles that are constrained by boundary conditions in space. In combination with the study of the effect of a spatially varying periodic lattice potential on these non-interacting fermions, the theory is very successful in explaining many aspects of the properties of crystalline materials, especially at low temperatures. For the discussion in the following sections it is important to review the key concepts and nomenclature used.

The first concepts to mention are those of lattice momentum and the Brillouin zone. Since the periodic potential of the lattice destroys the continuous translational symmetry of real space, momentum is no longer a good quantum number. However, a crystalline system still retains a discrete translational symmetry leading to the definition of a lattice momentum. For the purposes here, one can—with one important exception—treat this lattice momentum the same way as the normal momentum, and the explicit reference to the lattice will be omitted in the reminder of the thesis. The important exception is that the momentum space is finite and can be confined to the first Brillouin zone, the primitive unit cell of the reciprocal lattice related to the real lattice via a Fourier transform. Under these conditions it is possible to show that the resulting single-electron energy–momentum dispersion in a crystalline solid is given by a series of energy bands $\epsilon_N(\mathbf{k})$ in the Brillouin zone. Here ϵ is the energy, $\mathbf{k}$ the momentum vector and N the band index. If one neglects the Coulomb and magnetic interaction between the electrons then these energy levels are occupied at zero temperature according to Pauli's Exclusion Principle up to a maximum energy, the so called Fermi energy ϵ_F. Each energy level is doubly occupied due to the degeneracy in spin space. It can be shown that in the thermodynamic limit ϵ_F is independent of sample size. These concepts are illustrated in Fig. 2.1 for the example of copper [6, p. 289].

2.1 Itinerant Electron Systems

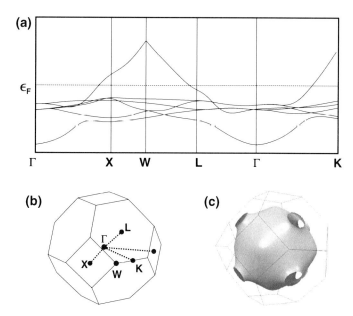

Fig. 2.1 **a** The band structure of Cu close to the Fermi energy ϵ_F is shown (graphic based on [6, p. 289]). The bands are plotted along lines in momentum space connecting points of high symmetry in the Brillouin zone. The position of these points is indicated in **b**, where the first Brillouin zone of Cu is given. The Fermi surface of Cu is shown in **c** [8, 9]

The wave vector in momentum space describing the Fermi surface is called the Fermi momentum $\mathbf{k}_F$. The Fermi surface topology is in general complicated. In particular, more than one band usually crosses the Fermi energy, giving a number of disconnected Fermi surfaces.

Most results of the theory take their simplest form for spherical Fermi surfaces. The magnitude of the Fermi momentum, k_F, is for example independent of direction and equal to the radius. For this reason mathematical results in this thesis are often presented for isotropic Fermi surfaces only. However, these are generally extendable to more complicated topologies.

The elementary excitation of such a system is the occupation of an unfilled level with an electron from a filled level. Whenever an electron crosses the Fermi energy a 'hole' is left behind in the states below the Fermi energy. The resulting symmetry in the number of particle and hole excitations is a fundamental property of a fermionic system as considered here. The typical width in energy of the electronic bands is of several eV. Therefore, any measurement of the system involving energy scales of not more than a few meV probes the physics of the energy–momentum dispersion close to the Fermi surface. As a result many properties of the system can be written in terms of a linearised band structure at ϵ_F. In connection to this an important quantity is the density of states $g(\epsilon)$, with $g(\epsilon)d\epsilon$ giving the number of electronic states in the energy range from ϵ to $\epsilon + d\epsilon$ normalised by the sample volume V.

An important class of measurements are thermodynamic experiments at low temperature. Since a temperature of 10 K corresponds to an average thermal energy of the order of 1 meV the above assumption of linearisation is often fulfilled experimentally. In this limit the electronic specific heat c_{el} is linear in temperature T and given by

$$c_{el} = \frac{\pi^2}{3} k_B^2 T g(\epsilon_F), \tag{2.1}$$

with k_B being Boltzmann's constant. The factor $(\pi^2/3)k_B^2 g(\epsilon_F)$ is called Sommerfeld coefficient γ. In the case of an isotropic surface, γ can be written as

$$\gamma = \frac{k_B^2}{3} k_F^2 \left(\left. \frac{\partial \epsilon}{\partial k} \right|_{k_F} \right)^{-1}. \tag{2.2}$$

Comparing this result to the Sommerfeld coefficient of the free electron gas,

$$\gamma_{gas} = \frac{m_e k_B^2}{3\hbar^2} k_F, \tag{2.3}$$

leads to the definition of the thermodynamic effective mass m^* as

$$m^* = \hbar^2 k_F \left(\left. \frac{\partial \epsilon}{\partial k} \right|_{k_F} \right)^{-1}. \tag{2.4}$$

A second class of experiments such as resistivity measurements or the response to strong magnetic fields involves dynamic properties of the electronic states. An appropriate theoretical framework for a discussion of most of the observed effects is the semiclassical theory. The basic components of this theory are electronic wavepackets in real space that represent an excitation of spin 1/2 and charge $-e$ being spread out over several lattice sites. If the analysis does not require too localised a wave function in real space, then one can represent it as a superposition of only a small number of momentum eigenstates around an average Fermi momentum $\mathbf{k}_F$. If one assumes furthermore that the applied electric or magnetic field is varying only slowly over the width of the wave packet, then it is possible to treat the field effectively as a classical force acting on the centre of the wave packet. The above described hierarchy of scales is shown schematically in Fig. 2.2.

An important variable in this theory is the (group) velocity of the wavepacket with which its centre of mass moves. It is given by

$$\mathbf{v}(\mathbf{k}) = \frac{1}{\hbar} \frac{\partial \epsilon}{\partial \mathbf{k}}, \tag{2.5}$$

2.1 Itinerant Electron Systems

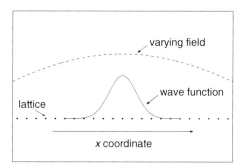

Fig. 2.2 This figure gives a overview of the hierarchy of scales in real space for which the semiclassical description is appropriate. The electron wavepacket described (*blue*) can only be localised to within several lattice site (*black dots*). Furthermore, the variations of the applied field have to be small on the scale of the wavepacket

with $\hbar$ being Planck's constant. The comparison of the momentum **p** of a wavepacket given by

$$\mathbf{p} = \hbar \mathbf{k}, \tag{2.6}$$

with the definition $\mathbf{p} = m^* \mathbf{v}$ for the classical momentum leading to a definition of an effective dynamic mass m^* as

$$m^* = \hbar^2 k_F \left(\left. \frac{\partial \epsilon}{\partial \mathbf{k}} \right|_{k_F} \right)^{-1}. \tag{2.7}$$

Note here that this is the same definition as has been obtained previously for the thermodynamic effective mass and no distinction between the two will be made in the following.

Within the semiclassical model one can derive the equations of motion for the position **r** and momentum **k** of a wavepacket under the influence of an electric field **E** and a magnetic field **H** as

$$\dot{\mathbf{r}} = \mathbf{v} = \frac{1}{\hbar} \left. \frac{\partial \epsilon}{\partial \mathbf{k}} \right|_{k_F}, \tag{2.8}$$

$$\hbar \dot{\mathbf{k}} = -e \left[\mathbf{E} + \frac{1}{c} \mathbf{v} \times \mathbf{H} \right]. \tag{2.9}$$

Here e is the charge of an electron and c is the velocity of light.

After having introduced the fundamental concepts of the theory of a non-interacting Fermi gas I will in the following discuss its results regarding the effects of magnetic and electric fields in more detail.

2.1.2 Magnetic Field Effects

Magnetic field effects are generally treated in two regimes. In the first, one assumes that the dynamics of electrons due to the applied field, i.e. electron orbital effects, are negligible. An example for this limit is Pauli paramagnetism. The second case takes full account of the orbital motion leading to such effects as quantum oscillations.

2.1.2.1 Pauli Paramagnetism and an Extension Thereof

If a magnetic field $\mu_0 H$ is applied to an itinerant paramagnetic electron system then each electron can orient its spin parallel ($\uparrow$) or antiparallel ($\downarrow$) to the magnetic field.

Since orbital motion of the electrons will be neglected here the only effect of the field is a shift of the energy levels for the spin-up and spin-down electrons by $\pm \mu_B H$. Therefore, in the presence of the magnetic field the density of states for each species of electrons can be given in terms of the zero field density of states $g(\epsilon)$ by

$$g_{\uparrow/\downarrow}(\epsilon) = \frac{1}{2} g(\epsilon \mp \mu_B \mu_0 H). \tag{2.10}$$

The requirement of conservation of number of electrons N in the system then defines implicitly the Fermi energy ϵ_F at zero temperature via

$$N = \int_{-\infty}^{\epsilon_F} (g_\uparrow + g_\downarrow) d\epsilon. \tag{2.11}$$

In the limit where the magnetic field is small, one can assume that the density of states is effectively constant, resulting in the Fermi energy ϵ_F being field independent. Furthermore, the magnetization M is given in terms of the number of spin up electrons, $n_\uparrow$, and spin-down electrons, $n_\downarrow$, by

$$M = -\mu_B(n_\uparrow - n_\downarrow) = -\mu_B(-\mu_B \mu_0 H g(\epsilon_F)), \tag{2.12}$$

resulting in the familiar expression for the susceptibility of Pauli paramagnetism χ of

$$\chi = \mu_B^2 g(\epsilon_F). \tag{2.13}$$

Two things should be mentioned. First of all the derivation at no point made any reference to the band structure other than the density of states, nor to the orientation of the magnetic field. Pauli paramagnetism is therefore isotropic. Furthermore, for simplicity, the argument was presented only for zero temperature, but it can be shown that χ is, to first order, temperature independent.

2.1 Itinerant Electron Systems

The reason the effect was presented in some detail here is because in the following it will be discussed what occurs when the magnetic field is sufficiently strong such that the assumption of the constant density of states breaks down (however orbital effects will still be neglected). This, for example, can happen at experimentally accessible fields if there is a peak in the density of states close to the zero field Fermi energy. This thought experiment is presented in Fig. 2.3.

In part (a), a possible situation in zero magnetic field ($H_0 = 0$) is shown. Here the density of states $g_{\uparrow/\downarrow}(\epsilon)$ are shown as a function of energy for the spin-up (blue) and spin-down (red) electrons. The density of states has a peak below the Fermi energy ϵ_F at the energies $\epsilon_{c\uparrow}/\epsilon_{c\downarrow}$ which are identical in zero field. If one now applies a magnetic field H_1 (part b) the energy of the spin-up band is lowered whereas the energy of the spin-down band is increased. However, whereas the difference in shift between the two bands is given by $\mu_B H$, the

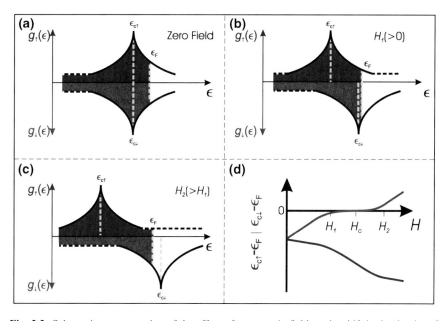

Fig. 2.3 Schematic representation of the effect of a magnetic field on the shift in the density of states of the spin up ($g_\uparrow(\epsilon)$) and spin down ($g_\downarrow(\epsilon)$) Fermi surfaces neglecting orbital effects. In zero magnetic field, as shown in **a**, the density of states of both spin species are the same. In the particular scenario here both density of states have a peak at $\epsilon_{c\downarrow}$ and $\epsilon_{c\uparrow}$ below the Fermi energy ϵ_F. The *filled states* of the two spin species are shown in *blue* (spin-up) and *red* (spin-down). In **b** and **c** it is shown that, when applying successively larger magnetic fields H_1 and H_2, the bands shift relative to each other with the density of states peak of one of the spin species being pushed through the Fermi energy. As discussed in the text the condition of overall number conservation causes the rate of change of the density of states to be a nonlinear function of applied field. This is shown schematically in **d** where the distance in energy of each of the density of states peaks from the Fermi energy ϵ_F, i.e. $\epsilon_{c\uparrow} - \epsilon_F$ and $\epsilon_{c\downarrow} - \epsilon_F$, is given as a function of magnetic field. As shown, $\epsilon_{c\uparrow} - \epsilon_F$ and $\epsilon_{c\downarrow} - \epsilon_F$ are not a linear function of $H - H_C$ where H_C is the critical field at which the peak in the density of states of one of the spin species crosses the Fermi energy

individual relative shift with respect to the Fermi energy is determined by number conservation. The consequence of this is that when the Fermi energy in one of the spin subspecies gets close to a peak in the density of states then the rate at which the Fermi energy approaches the peak as a function of applied magnetic field will slow down.

At a sufficiently high magnetic field the Fermi energy will have crossed the peak in the density of states as shown in Fig. 2.3c. Some careful remarks are necessary here. First, the above discussion completely ignores electron–electron interaction effects. Secondly, here the orbital effects of electronic states, leading for example to the phenomenon of quantum oscillations [10], were neglected. Overall one would not expect necessarily to be able to make quantitatively correct predictions based on this 'Gedankenexperiment'. However, one main qualitative conclusion is expected to hold. Close to a density of states peak one cannot expect that the distance in energy of the Fermi energy to that peak, i.e. $(\epsilon_C - \epsilon_F)$, is directly proportional to $(H - H_C)$, which is the difference of the magnetic field H to the critical magnetic field H_C at which the Fermi energy would lie at the maximum of the density of states peak. This is shown explicitly in Fig. 2.3d, where the relation between the two is shown for the particular scenario discussed here.

2.1.2.2 Quantum Oscillations

In the previous section it was assumed that the only coupling to the magnetic field is via the spin of the electrons. However, according to the semiclassical model introduced earlier, there exists a coupling between the magnetic field $\mathbf{H}$ and the momentum $\mathbf{k}$ of an electron wavepacket:

$$\hbar \dot{\mathbf{k}} = -e \left[\frac{1}{c} \mathbf{v} \times \mathbf{H} \right]. \quad (2.14)$$

On closer inspection of this equation one finds that the trajectory of the wavepacket in momentum space is along a path of constant energy in a plane perpendicular to the applied magnetic field $\mathbf{H}$. In other words, if the wavepacket originates at the Fermi surface its trajectory is a trace along the Fermi surface in a plane perpendicular to the magnetic field. In the following it will be assumed for simplicity that the plane in which the motion takes place in momentum space is the $k_x - k_y$ plane. One consequence of this is that k_x and k_y are not good quantum numbers any more.[1]

In suitable Fermi surface topologies the path of the electron will be closed, resulting in a motion in momentum space and real space that is periodic. Similarly

---

[1] They are replaced for a spherical Fermi surface by $(k_x \pm i k_y)$.

2.1 Itinerant Electron Systems

to the situation of electronic orbits in atomic physics it follows that the motion has to satisfy a quantisation condition which can be expressed in terms of the area a enclosed in momentum space and an integer n as

$$a = \left(n + \frac{1}{2}\right)\frac{2\pi eH}{c\hbar}. \tag{2.15}$$

The above relation is known in the literature as Onsager's relation [11]. A consequence of it is that the eigenstates in momentum space now are lying on tubes (called Landau tubes) whose principle axis is determined by the direction of the magnetic field.

Since the radius of these tubes and, as can be shown, their degeneracy, changes with magnetic field, one expects the properties of the electronic system to vary as a function of field as well. The energy of each tube associated with a definite quantum number n has to vary as a function of field, leading to the the Landau tubes crossing the Fermi energy one by one. This leads to an oscillatory behaviour in the properties of the material that, as can be derived from the quantisation condition, is periodic in $1/H$. A detailed derivation of that dependence is given for example in the book by Shoenberg [10]. Here, I will summarise the main results in the limit of high quantum numbers n.

The first quantity which has to be affected is the density of states. The oscillatory component $\tilde{g}(\epsilon_F)$ of the total density of states $g(\epsilon_F)$ at the Fermi energy ϵ_F as a function of magnetic field H is proportional to [10]

$$\tilde{g}(E_F) \propto \left(\frac{eH}{c\hbar}\right)^{1/2} m^* \sum_{p=1}^{\infty} \frac{1}{p^{1/2}} \cos\left(2\pi p \left(\frac{Ac\hbar}{2\pi eH} - \frac{1}{2}\right) \pm \frac{\pi}{4}\right). \tag{2.16}$$

Several important things have to be noted. First of all, A is an extremal cross sectional area of all possible cuts through the Fermi surface perpendicular to the applied magnetic field H. The contributions of all other cross sections interfere destructively. Secondly the total signal is a sum over all harmonics p of a fundamental oscillation in inverse magnetic field $1/H$. The frequencies F of quantum oscillations are therefore quoted in units of Tesla. Furthermore there is a direct proportionality between F and the area A that depends only on universal constants. By measuring the effect of these oscillations in the density of states on the entropy of the system (magnetothermal oscillations) or the resistivity (Shubnikov–de Haas effect) one can, therefore, obtain information about the size of the extremal areas of a Fermi surface. In particular if the Fermi surface is quasi-two-dimensional (i.e. a tube) the cross section and thereby the oscillation frequency should be proportional to $1/\cos(\Theta)$ with Θ being the angle between the applied magnetic field and the principal axis of the Fermi surface.

If the effect is observed in the longitudinal magnetisation $M_\parallel$ or AC magnetic susceptibility it is usually referred to as the de Haas–van Alphen effect. $M_\parallel$ can be shown to be proportional to

$$\tilde{M}_\parallel \propto \left(\frac{eH}{c\hbar}\right)^{3/2} \frac{m^*}{H} \sum_{p=1}^{\infty} \frac{1}{p^{1/2}} \sin\left(2\pi p \left(\frac{Ac\hbar}{2\pi eH} - \frac{1}{2}\right) \pm \frac{\pi}{4}\right) \quad (2.17)$$

and therefore contains equivalent information on the quantum oscillation frequencies and thereby on extremal Fermi surface orbits.

2.1.2.3 Effect of Temperature on Oscillations

At non-zero temperature the occupation of energy levels does not have a sharp discontinuity at ϵ_F but varies smoothly according to the Fermi-Dirac distribution $f(\epsilon, \mu)$ with μ being the chemical potential. Pippard [12] proposed a simple argument for the effect of temperature on quantum oscillations. He assumed that one can view the smooth Fermi–Dirac distribution as a superposition of a range of systems at zero temperature with slightly different chemical potentials. The overall measured oscillations can therefore be thought of as the phase coherent sum over all these systems. Mathematically, this is equivalent to a convolution of the zero temperature oscillations with the distribution function given by the derivative of the Fermi–Dirac distribution, i.e. $-df(\epsilon, \mu)/d\epsilon$. It can be shown that this effectively leads in frequency space to a multiplication of the amplitude of the frequency spectrum with a temperature and field dependent prefactor R_{LK} given by

$$R_{LK} = \frac{x}{\sinh(x)},$$

where

$$x = 2\pi^2 kTm^*c/e\hbar H. \quad (2.18)$$

This is called the Lifshitz–Kosevich (LK) temperature dependence of quantum oscillations. Since the functional form depends on the effective mass m^* one can extract this information from the measured temperature dependence of each of the observed frequencies in the spectrum. Quantum oscillations can therefore be used as a Fermi surface specific probe of the effective mass.

Effect of Impurities

At the end of this section a comment should be made on the impurity dependence of the oscillation amplitude. A crucial component of the theory is the quantisation condition due to closed orbits in momentum space. The presence of impurities means that an orbiting electron will be scattered at a rate τ that is related to the

2.1 Itinerant Electron Systems

mean free path l of the sample via the Fermi velocity $\mathbf{v}_F = l/\tau$. This is equivalent to a finite lifetime of the Landau levels and results in a phase smearing of the oscillations. It can be shown that the amplitude of oscillations due to this effect depends exponentially on the mean free path l and the magnetic field H. Quantum oscillations are therefore usually only seen in very pure single crystals and/or high magnetic fields. in particular since the dependence on mean free path l is exponential even small improvements in sample quality can significantly improve the signal.

2.1.3 Electric Transport

Empirically it is found that the low temperature resistivity ρ as a function of temperature T for most metals is described by

$$\rho = \rho_0 + AT^2, \qquad (2.19)$$

with ρ_0 being a temperature independent constant and A a material specific prefactor. From the semiclassical equations it would follow that the momentum $\mathbf{k}$ of an electron wavepacket is related to an applied electric field $\mathbf{E}$ via

$$\hbar \dot{\mathbf{k}} = -e\mathbf{E}. \qquad (2.20)$$

This would imply that the momentum of the electron would grow in time without bounds, implying infinite conductivity. Therefore, in the absence of any inelastic scattering processes, the resistivity would be zero. One obvious source for such processes is impurity scattering, which cause the temperature independent contribution ρ_0 of the resistivity. Since electron–electron scattering cannot take place in the non-interacting electron approximation the only other possibilities for inelastic scattering are electron–phonon processes. These indeed give a temperature dependent contribution, however it is found to generally be proportional to T^5 at low temperatures and not the dominant source of scattering in materials with strong electron–electron interactions. It is one of the failures of the non-interacting electron approximation not to be able to explain the T^2 dependence of the resistivity, since it is dominated by electron–electron processes that are explicitly excluded. I will return to this problem briefly in the context of Fermi liquid theory.

Before finishing this section a brief comment on the magnetoresistance in high magnetic fields should be given.[2] It follows from an analysis of the semiclassical model of a single Fermi surface[3] that, if one applies a magnetic field $\mathbf{H}$ perpendicular to an electric field $\mathbf{E}$, the induced current $\mathbf{j}$ is related to $\mathbf{E}$ via

---

[2] The high field limit here is the same as for quantum oscillations.
[3] The presented equations only hold for closed Fermi surfaces. Open orbit surfaces require a separate treatment discussed for example in [6].

$$\mathbf{E} = \rho_N \mathbf{j}, \qquad (2.21)$$

with ρ_N being the resistivity tensor of the Nth Fermi surface. ρ_N has the form

$$\rho_N = \begin{pmatrix} \rho & -RH & 0 \\ RH & \rho & 0 \\ 0 & 0 & 0 \end{pmatrix} \qquad (2.22)$$

if the magnetic field is applied the z-axis and the electric field along the x-axis. Here R is the Hall coefficient and ρ is the high field limit of the longitudinal resistivity. It can be shown that ρ saturates in this limit and becomes independent of magnetic field [6]. In a material with N Fermi surfaces each of them can be thought of as a separate conductance channel. Therefore, their resistance tensors ρ_n are added in parallel resulting in an overall resistance of the form

$$\rho = \left(\sum_{n=1}^{N} \rho_n^{-1} \right)^{-1}. \qquad (2.23)$$

2.1.4 The Fermi Liquid

All results presented in the previous sections were based on the gross simplification of neglecting electron–electron interactions. Nevertheless, they are found to agree with most of the experimentally observed behaviour of a wide range of metals. In this section a theory of the interacting Fermi liquid will be introduced that was originally developed by Landau in a series of papers [1, 2, 3]. It not only shows how the previous results are stable even with strong electron–electron interactions but also gives insight into some of the previously unresolved issues such as the temperature dependence of resistivity. As the name suggests, Landau's Fermi liquid theory describes the properties of a system of Fermions that are strongly interacting not dissimilar to the strong interactions of atoms in a liquid.[4]

The first important component of Landau's theory is what is now known as the adiabatic continuity principle. Here one assumes that the interactions are 'switched on' from zero and increased continuously to their final magnitude. In this way the eigenstates of the non-interacting problem will evolve in a continuous way to the eigenstates of the fully interacting problem. Therefore, there exists a one-to-one correspondence between the low-lying excitations of the non-interacting and the interacting system. Due to this correspondence it is possible to still label the interacting quantum states with the quantum numbers of the non-interacting eigenstates, keeping in particular the distinction between occupied and unoccupied

---

[4] Though I will discuss here the theory primarily in terms of electrons it should be pointed out that it applies to all interacting Fermion systems, such as, for example, liquid ^{3}He.

levels. This preserves the concept of the Fermi energy ϵ_F and the Fermi surface. However, even though the elementary excitations which empty an occupied level and fill an unoccupied one obey Fermi statistics, they are no longer single electron excitations but excitation of the whole Fermi liquid. These excitations are therefore often called quasiparticles. An important difference to single particle excitations is that quasiparticle excitations are not independent of each other. An elementary excitation of the Fermi liquid involves a redistribution of the charge density which is the source of a residual interaction between the quasiparticles. This causes the life time of quasiparticles to be finite. It is therefore not a priori clear that they constitute a good description of the system on the timescale of an experiment.

The second important component of Landau's Fermi liquid theory addresses this issue by analysing the lifetime of quasiparticles due to interactions. The main effect limiting the lifetime of a high-energy quasiparticle excitation of energy ϵ is the decay into excitations of lower energy under conservation of the symmetry in the number of particles and holes. It can be shown that the decay rate for a single excitation in the presence of an otherwise filled Fermi sea is proportional to $(\epsilon - \epsilon_F)^2$. Based on this it can be argued that there can always be found a temperature T below which the excitations are sufficiently long lived such that on the time scale of the experiment the quasiparticle picture is an appropriate description of the system. This phase space protection of the scattering rate between quasiparticles is also the origin for the T^2-term in the temperature dependence of the resistivity.

In summary, these two fundamental concepts of the theory show that at sufficiently low temperature the excitations of a Fermi liquid take the form of quasiparticles that obey Pauli's exclusion principle. Furthermore, the concepts of filled and empty states, and therefore of a Fermi surface are preserved. An important consequence is that any property whose derivation in the non-interacting picture only takes into account properties close to the Fermi surface should still hold in the framework of Fermi liquid theory.

What does change contrary to the non-interacting case is therefore not the functional form of most of the physical properties of the system but the absolute values and the range of external parameters, such as temperature, over which it is applicable. This explains why the theory of a Fermi gas is so successful in qualitatively describing properties of a wide range of crystalline solid state systems at low temperatures without taking full account of electron–electron interactions.

The enhancement of the properties of the system with respect to the non-interacting band structure values is phenomenologically described by so-called Landau Fermi liquid parameters. These encapsulate the effect of interactions on the observable quantities in a phenomenological way. They represent, in principle, the prefactors of an expansion of the momentum and spin dependent interaction energy in harmonics of the Fermi surface. For the case of a spherical Fermi surface this can be done explicitly as an expansion in spherical harmonics. Here one finds for the measured effective mass m^* an enhancement over the non-interacting band mass m_{band} given by

$$m^* = \left(1 + \frac{F_1^s}{3}\right) m_{band}. \tag{2.24}$$

Furthermore the magnetic susceptibility is enhanced as

$$\chi = \frac{m^*}{m_{band}} \frac{1}{1 + F_0^a} \chi_{band}, \tag{2.25}$$

with F_1^s and F_0^a being two Landau parameters for an isotropic Fermi surface such as, for example, in the case of liquid ^{3}He at low temperatures. The factor $1/(1 + F_0^a)$ is also often referred to as the Sommerfeld–Wilson ratio of a Fermi liquid. It describes the ratio between the spin susceptibility and the effective mass enhancement and can therefore be thought as a guide to the importance of magnetic quasiparticle interactions.

An interesting observation relating to the prefactor A of the temperature dependent part of the resistivity was made by Kadowaki and Woods [13]. When plotting for a range of materials A versus the square of the Sommerfeld coefficient γ^2 it was found that there is a positive linear correlation between them. The ratio A/γ^2 of a material is referred to as the Kadowaki–Woods ratio. It can indeed be theoretically justified that A is proportional to the square of the effective mass $(m^*)^2$ and therefore γ^2 for a single spherical Fermi surface (see [14] and references therein). However, when considering multiband materials where the different Fermi surfaces act as parallel conductance channels but contribute 'in series' to the specific heat, it becomes apparent that the Kadowaki–Woods ratio cannot hold as a general law when considering the overall material properties but at most for the specific contribution of a single part of the Fermi surface to the overall measurable quantities.

Last but not least, it should be remarked that the electron–electron interaction and therefore the renormalisation of the electronic properties in a real multi band material are often significantly different from band to band and can even vary within a band. This again only affects the quantitative aspect of the renormalisation but does not change the qualitative functional form of those properties, as long as the basic assumptions of Fermi liquid theory such as the adiabatic continuity are not violated.

2.1.5 Beyond the Fermi Liquid

In the previous section Landau's Fermi liquid theory was introduced. Its main implication was that the functional form of the theoretical predictions of the Fermi gas also hold for strongly interacting systems at low temperatures and small excitation energies. Just how robust Landau's Fermi liquid theory is can for example be seen in heavy Fermion systems, where the charge carriers have effective masses of the order of 100 electron masses. The seemingly wide range of applicability the success of Landau's Fermi liquid theory makes the materials in which it fails to describe particularly interesting, not in the least, to say it in Landau's own words (used in a different context), since *'especially [as] the brevity of life does not allow us the luxury of spending time on problems which will lead to no new results'* [15].

There are of course ample examples of non-Fermi liquid behaviour such as BCS superconductivity. Here, the fundamental excitations are not quasiparticles but Cooper pairs. In particular, the Kohn–Luttinger theorem [16] shows that any Fermi liquid at low enough temperatures will become superconducting. However, the theorem does not give a lower bound for that transition temperature and indeed for a number of metals no superconductivity has been observed to date.

From the scientific point of view the materials which attract particular interest are those that exhibit novel properties that do not readily agree with Landau's Fermi liquid theory or BCS superconductivity. Most notably falling into both these categories are the high temperature cuprate superconductors discovered in 1986 [4]. First these materials have an unusual superconducting state at low temperatures. Secondly the 'normal' metallic state above the superconducting transition has many characteristics that are not in agreement with a normal Fermi liquid description. However it should be noted here that this does not exclude the existence of quasiparticles in these materials. In particular, for special representatives of these compounds in which the normally present disorder due to carrier doping is suppressed, it was possible to observe the above described phenomenon of quantum oscillations [17, 18, 19]. There is an exciting range of other materials and concepts that show non-Fermi liquid physics such as the fractional quantum Hall effect systems [20] and Kondo lattice systems [21].

Particular interest in recent years has been caused by systems that can be tuned by an external parameter from Fermi liquid behaviour to non-Fermi liquid behaviour due to their closeness to a so-called quantum critical point. In the following I will introduce both from an experimental and theoretical point of view the concept of quantum criticality and its implications. This can be by no means exhaustive due to the large amount of research existing in the field and the mathematical complexity involved in the theory. More detailed accounts are given in several extensive reviews covering both experimental and theoretical aspects [21, 22, 23].

2.1.5.1 Quantum Criticality

Second order phase transitions at finite temperature usually occur between a high temperature disordered and a low temperature ordered phase. The order is characterised by an order parameter that starts to grow continuously in the ordered phase from zero at the transition temperature. A typical example is the transition from a paramagnetic to ferromagnetic phase with the order parameter being the magnetisation M. An important aspect of second order transitions is that coherent fluctuations in the order parameter diverge both in length and timescales on

Fig. 2.4 The figure shows a generic scenario of quantum criticality. For details see text

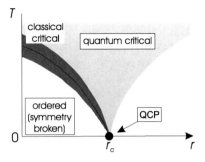

approaching the transition temperature from either side. In particular, the diverging time scale implies that the associated characteristic frequency ω and thereby the energy scale $\hbar\omega$ of these fluctuations goes towards zero. However, as soon as this energy scale is much smaller than that given by the temperature T, i.e. $\hbar\omega \ll k_B T$ they are effectively classical in nature. For classical fluctuations only the spatial dimensions have to be taken into account. Therefore, the critical fluctuations of any finite temperature second order phase transition fall into the universality classes of classical second order phase transitions.

However, if one can suppress the finite temperature transition towards $T = 0$ by using a tuning parameter such as pressure or magnetic field, then the region over which the critical fluctuations are classical in nature also shrinks to zero. This situation is shown schematically in Fig. 2.4. Here the second order transition line between the disordered and ordered state is shown in black with the region of classical fluctuations around the transition being indicated in red. When suppressing the second order phase transition towards zero temperature then the region over which the classical critical fluctuations dominate is reducing. If one crosses the transition at $T = 0$ as a function of the control parameter r then the critical fluctuations associated with the phase transition at r_C have a purely quantum mechanical nature, hence the name quantum critical point. It is here where the description of Fermi liquid theory generally breaks down.

Above r_C one generally observes a finite temperature cross-over towards a Fermi liquid like behaviour at low temperatures. The quantum critical behaviour is expected to become non-universal above sufficiently high temperatures.

This kind of physics has been especially well studied in heavy Fermion compounds, where often occurring ferromagnetically or antiferromagnetically ordered ground states can potentially be suppressed in temperature by the application of magnetic field or pressure.

Two examples are shown in Fig. 2.5. On the left hand side the phase diagram of YbRh$_2$Si$_2$ is shown in the $H - T$ plane ([25], graphic reproduced from [23]). Here the colour scheme shows the power law of the temperature dependence of the resistivity, with blue being 2 (Fermi liquid like behaviour) and the other end of the colour scale (orange) corresponding to 1. On applying a magnetic field one observes a suppression of the antiferromagnetic phase to below the experimentally accessible range. At the critical field the temperature dependence of the resistivity

2.1 Itinerant Electron Systems

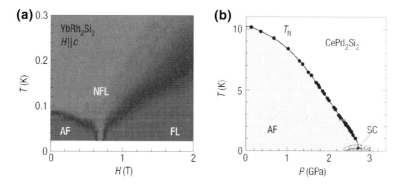

Fig. 2.5 Two examples of materials believed to be close to a quantum critical point (graphics reproduced from [23]). **a** The exponent of the temperature dependent part of resistivity for YbRh$_2$Si$_2$ as a function of temperature T and magnetic field H as derived from the logarithmic derivative of the resistivity with respect to temperature [25]. The colour scheme runs from 1 (*red*) to 2 (*blue*). The low field/low temperature phase is antiferromagnetic whereas the high field/low temperature phase is consistent with Fermi liquid like behaviour. In between the two states a quantum phase transition is expected to occur at the field where the measured resistivity is linear down to lowest temperatures. **b** The phase diagram of CePd$_2$Si$_2$ as a function of pressure. Again an antiferromagnetic phase transition is suppressed towards $T = 0$. However, at low temperatures a superconducting phase intervenes. The superconductivity is believed to be unconventional in that it is mediated by magnons rather than phonons

is non-Fermi liquid like over the whole range. At higher fields Fermi liquid behaviour in the resistivity is recovered.

In Fig. 2.5b the phase diagram of CePd$_2$Si$_2$ under pressure is shown [26] (Graphic reproduced from [23]). Here again the second order phase transition is suppressed towards zero temperature, this time by applying pressure.

In the second example a particularly significant aspect of quantum criticality is realised. What is often observed in the vicinity of quantum critical points is the stabilisation of novel phases with physical properties not usually observed. In CePd$_2$Si$_2$ for example the superconducting ground state is believed not to be due to electron–phonon interactions but to electron–magnon interactions [26]. This is particularly surprising since conventional superconducting states are destroyed by magnetic interactions.

The currently most successful approach in describing the quantum critical phenomenon was pioneered by Hertz [27] and Millis [28]. This theory is based on a renormalisation group approach similar to the theory of classical critical phenomena and leads to a classification of quantum phase transitions into certain universality classes. The fundamental difference is that for quantum phase transitions it is not only spatial but also temporal fluctuations that are included in the quantum mechanical action describing the system. Discussing Hertz–Millis theory in detail is beyond the scope of this brief introduction to the subject. However good introductions can be found in the aforementioned review articles [21, 22, 23]. Some of the results thought to apply to Sr$_3$Ru$_2$O$_7$, the system studied in this project, will be summarised in Sect. 2.2.2.

2.2 The Physics of the Ruthenate Family

Having discussed the general aspects of the physics of metals, this section will concentrate on the relevant experimental data and theoretical models specific to $Sr_3Ru_2O_7$, the material studied in this project, and Sr_2RuO_4, whose physical properties are of importance for the characterisation of the experimental setup developed for the measurements.

Both materials have generated significant scientific interest. Sr_2RuO_4 was the first layered perovskite material isostructural to the high-T_C cuprates that was found to be superconducting [29]. The transition temperature of 1.5 K is rather low compared to the high-T_C's, but it was proposed early on [30] to be one of a few superconducting materials whose order parameter has p-wave symmetry. A detailed presentation of the physics of Sr_2RuO_4 can be found for example in the review article by Mackenzie and Maeno [31].

$Sr_3Ru_2O_7$ is on the other hand in its zero-field ground state a highly enhanced paramagnet with a Wilson ration of ≈ 10 [32]. For comparison, the Wilson ratio of ^{3}He is 4 (see for example [33]). The interest in the material grew dramatically when the magnetic field phase diagram was studied in detail in a series of papers [34, 35, 36], reporting not only signatures consistent with a quantum critical point but also the existence of a novel phase in its vicinity that shows evidence of 'electron-nematic' properties in transport measurements.

2.2.1 Crystal Structure and Synthesis

The materials discussed in this thesis belong to the Ruddlesden–Popper strontium ruthenate series with the chemical formula $Sr_{n+1}Ru_nO_{3n+1}$, of which Sr_2RuO_4 and $Sr_3Ru_2O_7$ are the $n = 1$ and $n = 2$ members respectively. Their ideal crystal structures are shown in Fig. 2.6.

The fundamental building blocks are layers of corner-sharing ruthenium oxide octahedra. The number of adjacent RuO_2 layers is the reason for calling the materials the single-layer and bi-layer members of the series respectively. In between layers, Sr cations are arranged as shown.

The crystal structure of Sr_2RuO_4 with the tetragonal space group *I4/mmm* is faithful to the above ideal representation. The lattice parameters are $a = b = 3.862$ Å and $c = 12.722$ Å [38]. Contrary to that, the octahedra in $Sr_3Ru_2O_7$ show a small counter rotation of the octahedra by 7° [39]. The rotation direction alternates between the planes within a bi-layer. This structural difference causes a $\sqrt{2} \times \sqrt{2}$ reconstruction of the unit cell in the *ab*-plane and changes the space group of the material to the orthorhombic *Bbcb* with $a \approx b = 5.5$Å [39, 40]. This has significant implications for the electronic structure as will be shown further on.

The first single crystal samples of $Sr_3Ru_2O_7$ were grown by Cao et al. [41] with the flux-growth method. These samples were found to be ferromagnetic

2.2 The Physics of the Ruthenate Family

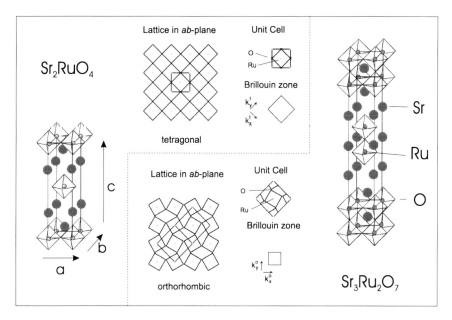

Fig. 2.6 Here the crystal structure is summarised for Sr_2RuO_4 and $Sr_3Ru_2O_7$ (graphics partially reproduced from [37]). Please see text for details

below 100 K. The material was subsequently synthesized in single crystal form by Ikeda et al. [42] by employing infra-red image furnace growth and found to be paramagnetic at low temperatures. The previously observed ferromagnetism was most probably due to inclusions of $Sr_4Ru_3O_{10}$ impurities. Since the image furnace grown crystal have a significantly improved quality they are referred to as second generation samples in the following. The thermodynamic properties, especially in magnetic field, were found to depend significantly on the sample quality and after a detailed growth study by Perry et al. [43] it was possible to reduce the residual resistivity of the materials by an order of magnitude (third generation samples). As will be seen in the following sections one cannot underestimate the importance of working with the highest purity samples available.

2.2.2 Thermodynamic Properties and Magnetic Phase Diagram

In the following I will introduce the phase diagram and the associated properties of the materials as established by previous measurements. I will in particular concentrate on thermodynamic measurements as well as resistivity before discussing the microscopic electronic structure in more detail in the next section.

2.2.2.1 $Sr_3Ru_2O_7$

Zero Field Low Temperature Properties

The crystal structure of $Sr_3Ru_2O_7$ is highly anisotropic, and this anisotropy is reflected in a number of properties such as its resistivity. In Fig. 2.7 the in-plane resistivity ρ_{ab} as well as the resistivity along the c-axis, ρ_c, are shown as a function of temperature T for second generation samples [32]. A pronounced asymmetry not only exists in the residual resistivity at zero temperature (observe the different scales for ρ_{ab} and ρ_c), but also in the A coefficient of the T^2 behaviour, as can be deduced from the gradient of the curves in the inset where the low temperature resistance is plotted against T^2.

The electronic specific heat c_{el} divided by temperature T is reproduced in Fig. 2.8a from [44] (data points not relating to the zero field properties have been

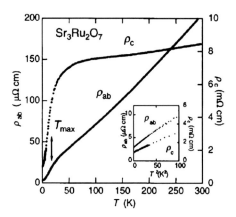

Fig. 2.7 In-plane and c-axis resistivity ρ_{ab} and ρ_c of $Sr_3Ru_2O_7$ as a function of temperature [32] (note the different scales for ρ_{ab} and ρ_c). The *inset* shows the same data in the low temperature range as a function of T^2

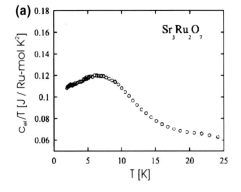
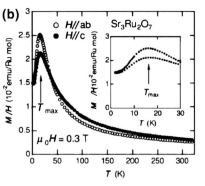

Fig. 2.8 a The electronic specific heat c_{el} of $Sr_3Ru_2O_7$ divided by temperature T as a function of T [44]. **b** The low field DC magnetic susceptibility of $Sr_3Ru_2O_7$ as a function of temperature T for the magnetic field applied along the crystallographic c-axis (*full circles*) and in the ab-plane (*open circles*) [32]. The *inset* shows a magnification of the low temperature range of the data

removed for clarity). c_{el}/T has a finite zero-temperature value corresponding to a Sommerfeld coefficient γ of 110 mJ/Ru-mol K^2, which is relatively large in the context of the ruthenates. Furthermore, there is a significant hump in the specific heat at a temperature of the order of 8 K that is in general associated with the metamagnetic feature discussed later on.

As mentioned before, Sr$_3$Ru$_2$O$_7$ is found in zero field to be a highly enhanced paramagnet. Figure 2.8b shows the DC magnetic susceptibility as determined by measuring the magnetisation M resulting from a small applied magnetic field H as a function of temperature T for two different orientations of H [32]. Here one can see a remarkably isotropic susceptibility at low temperatures even though the material itself is strongly anisotropic, suggesting that the Pauli term dominates the paramagnetism (see Sect. 2.1.2). From the above measurements a Wilson ration of $\approx$10 can be deduced.

Metamagnetism and Quantum Criticality

The unusually high Wilson ratio can be interpreted as Sr$_3$Ru$_2$O$_7$ being a paramagnet close to a magnetically ordered state. Indeed when applying a magnetic field one can observe a rapid superlinear rise in magnetisation at a well defined field. This phenomenological effect is called metamagnetism and is commonly encountered in local moment systems [45] undergoing spin-flip or spin-flop transitions. Figure 2.9a shows the magnetisation of Sr$_3$Ru$_2$O$_7$ as a function of applied magnetic field for several temperatures [44]. Here the field is applied in the crystallographic ab-plane. The metamagnetic feature at approximately $5T$ is clearly identifiable and is more defined towards lower temperatures.

Grigera et al. [46] studied the AC magnetic susceptibility systematically as a function of temperature as well as magnitude and direction of the applied magnetic field. They found that with the field applied in the ab-plane the metamagnetic cross-over turns into a first order phase transition as indicated by a dissipative out of phase component of the AC susceptibility. This first order transition line terminates in a critical end point at approximately 1.2 K. What caused particular interest was that the temperature of the critical end point could be suppressed as a function of angle of the applied magnetic field, leading to the possibility of a new class of quantum critical point—a so-called quantum critical end point [34]. The state of the phase diagram in second generation samples is summarised in Fig. 2.9b. Here, the position of the resulting sheet of first order transitions as a function of magnetic field, field direction and temperature is shown in green. The black line indicates the experimentally determined position of critical end points. The critical field moves towards higher values as the field is rotated towards the c-axis. At $B||c$ the critical endpoint is suppressed to below experimentally measurable temperatures. This particular orientation of magnetic field is the main focus of the project reported in this thesis.

The most precise early evidence for quantum criticality in this system was obtained by a detailed study of magnetotransport. The resistivity data were

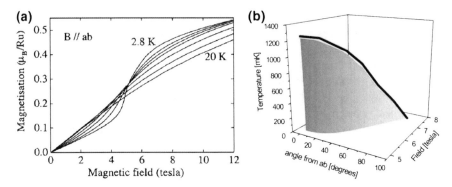

Fig. 2.9 a The magnetisation of $Sr_3Ru_2O_7$ as a function of applied magnetic field for a range of temperatures between 2.8 and 20 K [44]. The field is applied in the crystallographic *ab*-plane. The superlinear rise at low temperatures is the metamagnetic feature discussed in the text. **b** The *green* surface denotes the position of the metamagnetic first order transitions observed in second generation samples of $Sr_3Ru_2O_7$ as a function of field magnitude, field orientation (as angle between the magnetic field and the *ab*-plane) and temperature T [46]. The sheet of first order transitions terminates in a line of critical end points shown in *black*. For the field applied along the *c*-axis (90° on the angular scale) the data are consistent with the critical end point being suppressed to zero temperature

analysed at several magnetic fields H under the assumption that ρ is a combination of a temperature independent impurity scattering term, the residual resistivity ρ_{res}, and a temperature dependent power law contribution AT^α, i.e.

$$\rho(T,H) = \rho_{res}(H) + A(H)T^\alpha. \qquad (2.26)$$

In a Fermi liquid scenario α would be 2. Quantum critical theories however predict different power laws for the temperature dependence according to the universality class of the particular transition observed. A study by Millis et al. [47] analysed $Sr_3Ru_2O_7$ in particular and found that the temperature exponent should be 1 at the critical field H_C.

The main results of the experimental study are shown in Fig. 2.10 [34]. In panel (a) the extracted temperature exponent as a function of magnetic field[5] is shown. On approaching the critical field, the range in temperature over which the exponent of 2 consistent with Fermi liquid behaviour is observable, shrinks to below the lowest measured temperature. At the critical field a close to linear temperature behaviour is observed whereas at higher magnetic fields Fermi liquid like behaviour is recovered.

In panel (b) both the extracted residual resistivity ρ_{res} as well as the A coefficients extracted at lowest temperatures are plotted as a function of field in the vicinity of the critical field at $H_C \approx 8\,\text{T}$. The increased A coefficient in the vicinity

---

[5] To be more precise it is the logarithmic derivative of the temperature dependent part of the resistivity ρ with respect to temperature T, $\partial \ln(\rho - \rho_0)/\partial \ln T$, that is shown. Here ρ_0 is the residual resistivity. For more details see [34].

2.2 The Physics of the Ruthenate Family

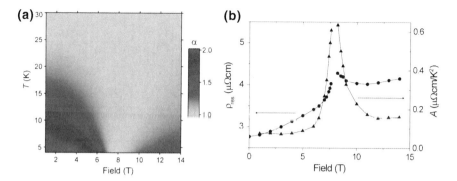

Fig. 2.10 a The logarithmic derivative of the temperature dependent part of the resistivity with respect to temperature as a colour scale in the magnetic field and temperature plane [34]. In the case of a pure power law AT^α this is equal to α. **b** The residual resistivity ρ_0 as well as the coefficient A of the temperature dependent part of the resistivity as a function of magnetic field and temperature (note the different scales of ρ_0 and A). For details of the parameter extraction from the original data please refer to [34]

of H_C indicates, in conjunction with the Kadowaki–Woods ratio (see Sect. 2.1.4), an increase in the effective quasiparticle mass of at least a fraction of the Fermi liquid excitations. However, a detailed theoretical understanding of this ratio in a realistic multiband system in a magnetic field is not developed completely. Therefore, a thermodynamic study of the specific heat is necessary to address the issue, in particular since the quantum critical scenario makes specific predictions about the temperature and field dependence of the specific heat [47, 48]. Furthermore the peak of residual resistivity at the critical field seems to suggest that the increase of the quasiparticle mass is accompanied by an increase of the impurity scattering. In hindsight this feature is most probably associated with the phase formation observed in the high purity third generation samples as discussed below.

Experimental evidence consistent with a quantum critical scenario was observed in the measurement of the electronic specific heat c_{el} close to the critical field at temperatures above 1.5 K by Perry et al. [44]. The results are shown in Fig. 2.11. Here c_{el} divided by temperature T is shown as a function of T for a range of magnetic fields. The field direction is parallel to the crystallographic c-axis. In particular, the functional form of the data at 7.7 T was found to be consistent with c_{el}/T having a singular component that diverges logarithmically at $T=0$ consistent with a quantum critical scenario [47].

Novel Quantum Phase and Nematic-Like Transport

The possibility of studying quantum criticality in a clean system motivated a detailed study into the optimisation of the crystal growth parameters by Perry et al. [43]. In the resulting third generation samples, mean free paths of the order of

Fig. 2.11 Electronic specific heat c_{el} divided by temperature T as a function of T. Measurements have been done in zero field, 7.7 T and 9 T with the field oriented parallel to the crystallographic c-axis [44]

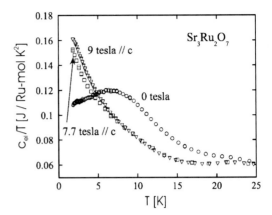

3,000 Å were observed. In these samples it was found by Grigera et al. [35] that in the vicinity of the proposed quantum critical end point a novel phase is stabilised which consequently is strongly disorder sensitive.

Figure 2.12 presents the experimental evidence for this phase as seen in electric transport and several thermodynamic measurements [35]. Part (a) shows the resistivity as a function of field for several temperatures. The resistance curves describe a distinct region as a function of magnetic field between 7.8 T and 8.1 T in which the resistivity is significantly increased and a much weaker temperature dependence observed compared to the surrounding states.

Experimental evidence that these features indeed mark first order phase transitions is shown in part (b). Here both the in-phase and out of phase components of the AC magnetic susceptibility[6] χ' and χ'' as well as the linear magnetostriction λ are shown as a function of magnetic field H for a sample temperature of 100 mK. The features (1) and (2) coincide with the boundaries of the anomalous resistive region in the previous graph. Furthermore the clear signatures in the loss term χ'' indicate that these are first order transitions. The width of the transitions as seen here is 50 mT. Feature (3) is a metamagnetic cross-over that is much wider in magnetic field and does not show up in χ''. In the inset of Fig. 2.12b a trace of magnetisation M as a function of temperature T is shown at the critical field H_C. Here a clear change in temperature dependence at 1.2 K is an indication for a thermodynamic transition into the new phase. However, further definite thermodynamic proof is needed to establish its exact nature.

Finally in Fig. 2.13 the magnetisation of ultra clean $Sr_3Ru_2O_7$ as a function of applied magnetic field is shown (data provided by R.S. Perry). The field orientation is parallel to the crystallographic c-axis and the sample temperature is 70 mK. The magnetisation is raising very fast at the transitions (1) and (2) and the crossover (3). The labels here correspond to the same used in Fig. 2.12.

---

[6] Strictly speaking AC magnetic susceptibility is not a thermodynamic probe but frequency dependent.

2.2 The Physics of the Ruthenate Family

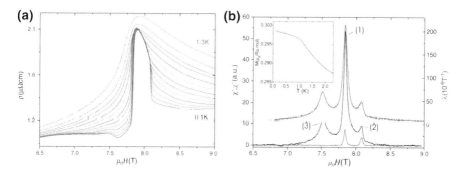

Fig. 2.12 a The in-plane resistivity of a third generation $Sr_3Ru_2O_7$ sample as a function of magnetic field H for several temperatures between 0.1 and 1.2 K [35]. The field is applied along the crystallographic c-axis. **b** Measurements of the components χ' and χ'' of the complex AC magnetic susceptibility as well as the linear magnetostriction λ as a function of magnetic field. Here the sample temperature is 100 mK and the field is applied along the c-axis. The *inset* shows the magnetisation of $Sr_3Ru_2O_7$ as a function of temperature at $\approx$8T, a field in between the two first order phase transitions that are observed in the other thermodynamic quantities [35]. Furthermore the two features coinciding with the resistive boundaries in **a** are labelled (*1*) and (*2*)

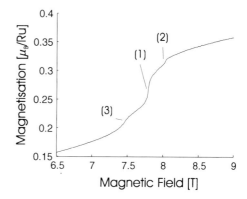

Fig. 2.13 Magnetisation of third generation $Sr_3Ru_2O_7$ samples as a function of applied magnetic field. The field orientation is parallel to the crystallographic c-axis and the sample temperature is 70 mK. The labels (*1*)–(*3*) are explained in the text (data provided by R.S. Perry)

That the novel phase between the boundaries (1) and (2) has indeed highly unusual resistive properties has been established by Borzi et al. [36] in a careful study of the resistivity as a function of not only applied field H and temperature T but also of the angle between the current I and H. In particular Fig. 2.9a shows the resistivity as a function of magnetic field H with the field applied at 77° from the crystallographic ab-plane. This field can be thought of as a linear combination of a field H_c parallel to the c-axis and a small added component H_a along the a-axis. Here, the label a has been chosen instead of b arbitrarily since a and b are interchangeable. The current I can now be applied either along ($I_\parallel$—shown in black) or perpendicular ($I_\perp$—shown in red) to the in plane field component H_{ab} as shown in the inset in Fig. 2.14a. While the resulting curves outside the new phase are indistinguishable, their behaviour inside the phase is dramatically different.

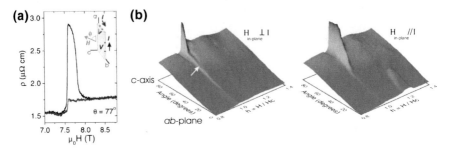

Fig. 2.14 a The resistivity ρ as a function of field H with the field applied at 77° from the ab-plane (i.e. close to the c-axis) [36]. The two traces correspond to the current being parallel to the in-plane component of the field (*black*) and perpendicular to it (*red*) as shown in the inset. **b** The in plane resistivity as a function of orientation and magnitude H of magnetic field. The two graphs correspond to the two different orientations of the current I with respect to the in-plane component of the magnetic field H as indicated. Please note that the field scale is given in H/H_C. Here H_C is the angle dependent position of the main first order metamagnetic transition. Furthermore, on the angular scale 90° corresponds to the magnetic field being applied parallel to the c-axis and 0° to the magnetic field being in the ab-plane. The *white arrow* indicates the point at which the two first order transitions enclosing the novel quantum phase merge, within experimental resolution

Anisotropic transport in itself is not unusual. Here, however, first of all it has been found in an accompanying neutron study that while the transport is anisotropic one cannot distinguish between the values of a and b within experimental resolution even under an applied in-plane magnetic field. Instead the data indicate that the symmetry breaking revealed by transport is intrinsic and the small applied in-plane magnetic field is merely 'aligning' different regions of domains of an intrinsically symmetry broken phase. In combination the results by Borzi et al. are consistent with the electronic liquid breaking the (discrete) rotational symmetry of the lattice—a behaviour called, in analogy to nematic phases in liquid crystals, '*electronic nematicity*'.

The full angular dependence for $I_\parallel$ and $I_\perp$ as a function of magnitude of applied field and orientation of the field relative to the ab-plane is shown in Fig. 2.14b. Please note that the field magnitude axis is divided by the critical field H_C which is angle dependent as has been discussed in the previous section. In addition to the anomalous phase close to the c-axis (at an angle of 90°) a further anomalous region can be identified in the ab-plane.

Finally, before ending this section, the most relevant information on the position of phase transition lines and the observance of 'electron-nematic'-like behaviour is summarised in Fig. 2.15. Panel (a) is reproduced from [49] and shows as a function of temperature, magnetic field and orientation of the field the position of the established first order transition lines in green whereas the regions under the blue domes are those in which the anomalous 'nematic' transport characteristics have been observed. In Fig. 2.15b the phase diagram in the c-axis as established by magnetisation M, AC magnetic susceptibility χ, linear magnetostriction λ and thermal expansion coefficient α is shown [35, 50]. The position of the first order

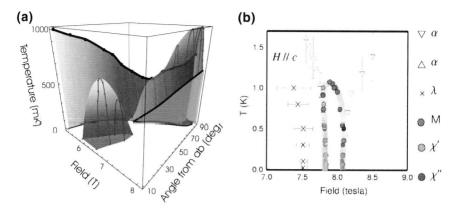

Fig. 2.15 Here the most relevant phase diagram information of high purity single crystal $Sr_3Ru_2O_7$ based on the current experimental data is shown. **a** The observed first order metamagnetic transition sheets as a function of temperature, magnetic field magnitude and orientation are shown in green. The *black lines* represent the lines of critical end points of these sheets. Furthermore the regions inside the *blue domes* have been found to show 'nematic-like' transport features as discussed in the text. On the angular scale 90° corresponds to the magnetic field being applied parallel to the c-axis and 0° to the magnetic field being in the ab-plane (figure reproduced from [49]). In **b** the phase diagram with the field applied parallel to the c-axis is shown. The *green lines* show the position of first order transitions whereas the blue line indicates the position of the putative second order phase transition. The other features are thermodynamic crossovers. The data points shown are extracted from DC magnetisation M, the components of the complex AC magnetic susceptibility χ, linear magnetostriction λ and thermal expansion coefficient α [35, 50]

transitions is marked in green as in the previous graph. The true thermodynamic nature of the 'roof' connecting the two first order transitions cannot be determined on the basis of the evidence that existed before the current project began. The other features indicated in this phase diagram are crossovers.

2.2.2.2 Sr_2RuO_4

In this section the thermodynamic properties of Sr_2RuO_4 will only be discussed insofar as they are relevant to the characterisation of the experimental setup developed in this thesis.

In zero field, Sr_2RuO_4 has a superconducting transition at $T_C = 1.5$ K. In Fig. 2.16a the electronic specific heat c_{el} divided by temperature T is shown as a function of temperature [51]. The sharp transition at T_C is clearly identifiable. Furthermore c/T above T_C is consistent with Fermi liquid theory. The magnitude of the Sommerfeld coefficient γ of the normal state is 38 mJ/Ru-mol K^2, approximately a third of the Sommerfeld coefficient observed in $Sr_3Ru_2O_7$. The results for electronic specific heat are usually quoted 'per mole ruthenium' (Ru-mol) since the valence electrons in these systems originate from the Ru $4d$-shell as will be discussed in the next section.

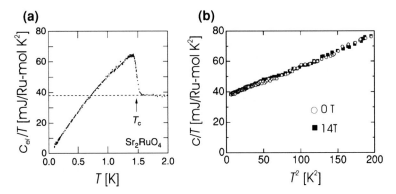

Fig. 2.16 a The electronic specific heat c_{el} of Sr_2RuO_4 divided by temperature T as a function of T [51]. The *dotted line* represents the normal state specific heat that would be observed if the transition to superconductivity at T_C would not take place. **b** The total specific heat c of Sr_2RuO_4 divided by temperature T as a function of T^2 with the sample being in zero magnetic field and in a field of $14T$ [52]

The second important feature for this project is that the specific heat of the normal state of this material is virtually independent of magnetic field. In order to illustrate this, Fig. 2.16b shows the specific heat divided by temperature as a function of T^2, for no magnetic field applied (circles) and with a magnetic field of 14 T applied (squares) [52]. The linear dependence on T^2 is caused by the phononic contribution to the specific heat.

2.2.3 Electronic Structure Properties

In the previous section the macroscopic properties of $Sr_3Ru_2O_7$ and Sr_2RuO_4 were presented. In particular it was shown that in certain regions of the phase diagram their properties are consistent with Fermi liquid theory. In this section the current experimental knowledge of the microscopic electronic structure, that is, the Fermi surface and band structure of these materials will be briefly reviewed. The section will begin with Sr_2RuO_4 since the more complicated band structure of $Sr_3Ru_2O_7$ can be understood more easily by starting from the band structure of Sr_2RuO_4.

2.2.3.1 Fermi Surface of Sr_2RuO_4

Sr_2RuO_4 is a material whose Fermi surface has been studied in great detail using quantum oscillations [54, 55]. The upper part of Fig. 2.17 shows as an example typical de Haas–van Alphen oscillations as a function of the magnetic field [53]. Due to the unusually high signal to noise ratio it is possible to observe the fundamental frequencies as well as higher harmonics as is shown in the frequency

2.2 The Physics of the Ruthenate Family

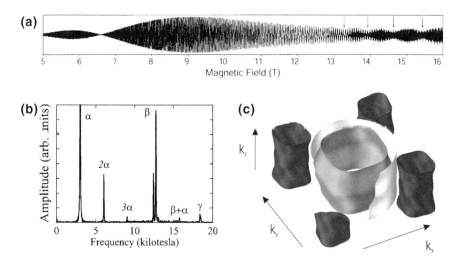

Fig. 2.17 a A typical trace of Sr_2RuO_4 quantum oscillations as a function of magnetic field observed in the de Haas–van Alphen effect [53]. **b** The frequency spectrum of quantum oscillation in inverse magnetic field for Sr_2RuO_4 [52]. The magnetic field is applied along the crystallographic c-axis. The fundamentals α, β and γ are labelled together with their harmonics and sum frequencies. **c** The Fermi surface of Sr_2RuO_4 as reconstructed from quantum oscillations in combination with LDA calculations [31]. The central surface is the β sheet surrounded by the γ surface. The α surface is centred on the corner of the Brillouin zone and shown four times here. The Fermi surface dispersions along k_z have been amplified by a factor 15 for reasons of presentation

spectrum in Fig. 2.17. In a detailed angular dependence study, Bergemann et al. [55] were able to reconstruct the Fermi surface of Sr_2RuO_4 as shown in Fig. 2.17c. It consists of three quasi-two-dimensional parts, the α, β and γ surfaces, that are relatively weakly warped in the k_z direction. The quasi-two-dimensionality can be readily understood by taking the crystal structure into account. The electronic states at the Fermi surface are derived from Ru $4d$ orbitals. Since these are arranged in layers of RuO octahedra, the hopping integral in a tight binding approach for electronic states close to the Fermi energy is significantly larger within a plane than in between planes. This leads to a very weak dispersion of the energy bands in the k_z direction of momentum space resulting in the observed quasi-two-dimensionality of the Fermi surface.

The quantum oscillation findings have been largely confirmed by ARPES measurements [56] and LDA band structure calculations [57], as shown in Fig. 2.18a and b, respectively.

ARPES and quantum oscillation measurements in $Sr_{2-x}La_xRuO_4$ samples, which are effectively electron-doped, made a definite identification of the α sheet as being of 'hole'-like character and the β and γ sheets as being 'electron'-like [58, 59].

Since the surfaces are quasi-two-dimensional the contribution of a single Fermi surface sheet of effective mass m^* to the Sommerfeld coefficient γ can be shown in the case of Sr_2RuO_4 to be [54]

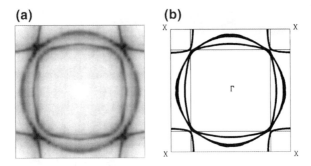

Fig. 2.18 a The Sr_2RuO_4 Fermi surface cross section in the k_x–k_y plane of the Brillouin zone as measured by ARPES (graphic reproduced from [31] based on [56]). **b** The same cross section as obtained by LDA calculations [57]. Here the thickness of the lines is indicating the warping of the surfaces in the k_z-direction. Γ labels the zone boundary whereas X labels the zone corner

$$\gamma = \frac{\pi N_A k_B^2 a^2}{3\hbar^2} m^*. \tag{2.27}$$

Here, N_A is Avogadro's constant, k_B Boltzmann's constant, $\hbar$ Planck's constant and a the lattice parameter of Sr_2RuO_4. By applying this equation to the measured masses from the de Haas–van Alphen experiments one deduces a total Sommerfeld coefficient of 38.5 mJ/Ru-mol K^2, in good agreement with the value obtained from bulk specific heat measurements.

A further test for the Fermi liquid description of the material is, if the Fermi surface volume can successfully account for all the valence electrons in accordance with Luttinger's theorem [60]. Using the experimental results one obtains a total of 4.05 electrons in the three observed valence bands, in agreement with the four remaining electrons in the Ru^{4+} state of the material.

Overall, one can conclude that the band structure of the normal state of Sr_2RuO_4 can be very well understood in terms of Fermi liquid theory.

2.2.3.2 Fermi Surface of $Sr_3Ru_2O_7$

The Fermi surface structure of Sr_2RuO_4 was discussed in some detail in the previous section because it is a good starting point for understanding the rather more complex Fermi surface structure in $Sr_3Ru_2O_7$.

The first basic assumption for the following line of argument is that in both materials it is the energy bands largely originating from the Ru 4d orbitals which cross the Fermi energy. The second assumption is that the change from single layer to double layer material causes a so-called bilayer splitting, i.e. doubling of the number of bands, but does not affect the quasi-two-dimensionality of the Fermi surface. The third assumption is that the rotation of the octahedra in the ab plane causes a $\sqrt{2} \times \sqrt{2}$ reconstruction of the unit cell but is otherwise, due to its smallness, only a weak perturbation.

2.2 The Physics of the Ruthenate Family

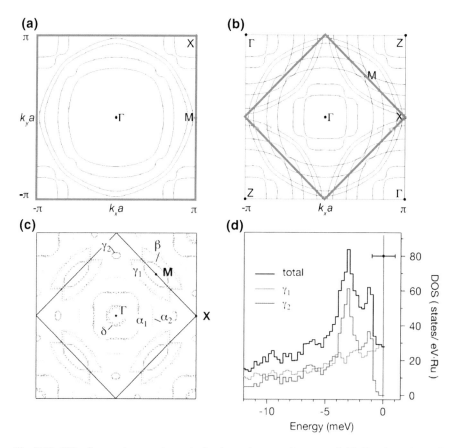

Fig. 2.19 This figure shows schematically how the complex zero field Fermi surface of $Sr_3Ru_2O_7$ can be understood by considering the effects of the differences in structure between the two materials being treatable as perturbations to the band structure. **a** The expected effect of a bi-layer splitting on the Fermi surface. The quasi-two-dimensionality of the Fermi surface is assumed to be retained. If one takes into account the $\sqrt{2} \times \sqrt{2}$ reconstruction of the unit cell due to the rotation of the RuO-octahedra one expects the Brillouin zone to halve and rotate by 45°. This new Brillouin zone is shown in *blue* in **b**. The bands have to be folded into this new zone as indicated. This results in a high degeneracy in the band structure along certain paths in momentum space. These are lifted, leading to a reconstruction of the Fermi surface. **c** The band structure in the k_x–k_y plane as extracted from an ARPES measurement [61]. For details regarding the labelling please see text. **d** The density of states as function of energy relative to the Fermi energy as measured by ARPES for the two γ pockets and their total sum. The error bar in energy represents the experimental uncertainty in the position of the Fermi energy. The plot is reproduced from [61] (labelling changed for consistency)

The effect of these assumptions on the Fermi surface are sketched in Fig. 2.19a and b which are reproduced from [37]. Part (a) shows the k_x–k_y plane of the Sr_2RuO_4 Brillouin zone. Here the previously presented Fermi surfaces are shown doubled due to the bilayer splitting. In part (b) the new Brillouin zone due to the

$\sqrt{2} \times \sqrt{2}$ reconstruction of the unit cell is shown in blue. The bands are 'backfolded' resulting in the shown complex Fermi surface topology.

If one would stop the 'Gedanken experiment' here, then the eigenenergies of the resulting bands would be degenerate along several paths in momentum space. Perturbation theory predicts that in a material these degeneracies are lifted. This in turn would lead to a reconnection of the Fermi surface which in its details is sensitive to the exact prevailing energetics in the material. Figure 2.19c shows the Fermi surface in the k_x–k_y plane as extracted from an ARPES measurements by Tamai et al. [61]. Though different in detail, the overall geometrical similarities between the measurement and the prediction based on a very simple model is remarkable. One can therefore assume as an hypothesis that the above assumptions are a good starting point for an understanding of the band structure of $Sr_3Ru_2O_7$ in zero field. The Fermi surfaces in Fig. 2.19c are therefore labelled in accordance to their relation to the Fermi surface and band structure in Sr_2RuO_4. For example the α_1 and α_2 sheets are mainly derived from the α Fermi surface sheet in Sr_2RuO_4. One obvious difference to the Fermi surface as expected from the simple model presented earlier is the additional δ-surface centred at the Γ point of the Brillouin zone. LDA calculations by Singh [61] indicate that this pocket is derived from a band that did not cross the Fermi energy in Sr_2RuO_4 but also has primarily a Ru $4d$ orbital character (in this case $d_{x^2-y^2}$). Overall, the ARPES measurement observed five sheets that certainly cross the Fermi energy, namely α_1, α_2, β, γ_1 and δ. LDA calculations indicate that δ is indeed bilayer split but that the splitting could not be resolved. The sixth Fermi surface shown, γ_2 is particular in that the energy resolution of the ARPES measurement is not sufficient in its own in order to decide if this pocket is indeed crossing the Fermi energy or is situated just below it. However, the γ_2 surface proves to be critical to the properties of the material. First of all if one excludes it then the observed Fermi surface can only account for $\approx$40% of the total measured specific heat, as will be discussed in more detail below. Secondly ARPES measurements of the band structure below the Fermi energy revealed two peaks in the density of states $g(\epsilon)$ of the γ_2 pocket. This is shown in Fig. 2.19d (the graph is reproduced from [61] with alterations to the labelling for consistency). Here the density of states (DOS) of the two γ pockets and the total sum is shown as a function of energy relative to the Fermi energy. A small peak is situated at ≈ -1 meV and a stronger one related to a saddle point in the band structure at ≈ -4 meV. The error bar shows the uncertainty in the determination of the position of the Fermi energy in the experiment.

The detailed properties of these six Fermi surfaces will be presented at the end of this section together with the results from de Haas–van Alphen measurements. Quantum oscillations in resistivity (Shubnikov–de Haas effect) and magnetic susceptibility (de Haas–van Alphen measurements) in $Sr_3Ru_2O_7$ have been reported in several papers [62, 63]. The most detailed study on the samples with the highest purity has been performed recently by Mercure et al. [37]. On the low field side, with magnetic field applied parallel to the crystallographic c-axis, they were able to observe the same five Fermi surface sheets as the ARPES measurements, namely α_1,

2.2 The Physics of the Ruthenate Family

Table 2.1 The properties of the $Sr_3Ru_2O_7$ Fermi surfaces as measured by de Haas–van Alphen (dHvA) and ARPES experiments

	α_1	α_2	β	γ_1	γ_2	δ
dHvA						
F (kT)	1.78	4.13	0.15	0.91	–	0.43
Area (% BZ)	13.0	30.1	1.09	6.64	–	3.14
m^* (m_e)	6.9 ± 0.1	10.1 ± 0.1	5.6 ± 0.3	7.7 ± 0.3	–	8.4 ± 0.7
ARPES						
Area (% BZ)	14.1	31.5	2.6	8.0	<1	2.1
m^* (m_e)	8.6 ± 3	18 ± 8	4.3 ± 2	9.6 ± 3	10 ± 4	8.6 ± 3

F is the observed frequency in $1/H$ of quantum oscillations, BZ stands for Brillouin zone and m^* is the effective mass in units of electron mass m_e. Reproduced from [37]

α_2, β, γ_1 and δ. The angular dependence of the oscillations is consistent with these Fermi surface pockets being quasi-two-dimensional.

The amplitude of the extremely small γ_2 pocket was not identified in the initial de Haas–van Alphen measurements performed in the study. For the other five pockets both the cross-sectional area and the effective mass were measured with relatively high accuracy. Table 2.1 reproduced from [37] gives the detailed results as obtained from both measurements

The overall agreement between the two measurements is good with the de Haas–van Alphen results showing the smaller experimental uncertainty. An important test is if these measurements can explain the experimentally observed bulk specific heat. If one takes only the five Fermi surfaces excluding γ_2 into account than the expected Sommerfeld coefficient from the de Haas–van Alphen experiment is (56 ± 1) mJ/Ru-mol K^2 and from the ARPES measurements (67 ± 12) mJ/Ru-mol K^2, both of which account for only approximately half of the measured specific heat value of 110 mJ/Ru-mol K^2.

The γ_2 pocket on the other hand occurs four times in the Brillouin zone and furthermore is probably bilayer split. The contribution to the Sommerfeld coefficient is therefore eight times the value as calculated based on its effective mass m^* alone. Taking the mass of the γ_2 pocket as measured by ARPES into account would therefore result in revised estimates of the Sommerfeld coefficient of (127 ± 27) mJ/Ru-mol K^2 (ARPES) and (116 ± 16) mJ/Ru-mol K^2 (de Haas–van Alphen). These values would be in broad agreement with the specific heat measurements. Since the size of the γ_2 is negligible on the scale of the size of the Brillouin zone it has a very small effect on the Luttinger sum for the total number of electrons enclosed by the Fermi surfaces.

Contrary to ARPES, the de Haas–van Alphen effect is intrinsically suited to extend the Fermi surface study to high magnetic fields. Here I will summarize the main findings by Mercure et al. [37] as relevant to this thesis.

1. One of the most important results is the absence of any significant effective mass increase in the vicinity of the critical field for any of the five observed frequencies. Figure 2.20a shows this as an example for three of the observed

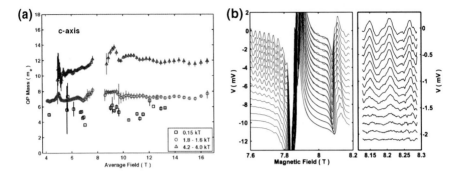

Fig. 2.20 a The quasiparticle mass as a function of magnetic field for some of the frequencies observed by de Haas–van Alphen measurements. **b** Data of second harmonic de Haas–van Alphen measurements. The traces which correspond to different sample temperatures have been offset for clarity. In the *right panel* the data inside the novel quantum phase are shown after the background given in *red* in the *left panel* has been subtracted (graphics courtesy of J.-F. Mercure). For more details see [37]

frequencies. Here the effective quasiparticle (QP) mass is plotted as a function of magnetic field applied parallel to the c-axis of $Sr_3Ru_2O_7$. Therefore, the specific heat contribution of the five Fermi surface sheets α_1, α_2, β, γ_1 and δ cannot explain the observed peak in the A-coefficient of resistivity as discussed in the previous section.

2. Mercure et al. succeeded in measuring quantum oscillations in the region of the new phase as shown in Fig. 2.20b. The left panel shows the second harmonic of an AC magnetic susceptibility measurement as a function of magnetic field for several temperatures. Here the traces have been offset in order of temperature for clarity and the upmost trace corresponds to the lowest temperature. The two strong features correspond to the phase boundaries of the novel phase and the red curves are a polynomial background fit to the data. The right panel shows the data with this background being subtracted. Here the quantum oscillations are clearly identifiable. The phase of the oscillations has no apparent temperature dependence. Furthermore the amplitude dependence on temperature is well described by the Lifshitz–Kosevich formula. One can conclude that at least part of the excitation spectrum in the novel quantum phase is therefore consistent with the existence of Fermi liquid quasiparticles. That does not mean however that all electronic degrees of freedom can necessarily be described as itinerant quasiparticles, with other examples where this is considered to be a possibility being high-T_C superconductors [64] and heavy Fermion Kondo systems [22].

3. The observed frequency structure in the high field Fermi liquid state is highly fractured and the amplitude significantly suppressed. Possible reasons for this can be magnetic breakdown and Fermi surface reconstruction.

The ARPES and de Haas–van Alphen measurements presented here therefore seem to indicate that the physics of $Sr_3Ru_2O_7$ is driven by a single band of the complicated band structure which is associated with the γ_2 pocket. It has to be emphasized that this conclusion is reached by elimination of the remaining Fermi surface and the evidence for a peak in the single-particle density of states associated with γ_2. No direct evidence of the field dependence of this peak or the effective mass of γ_2 exists.

Within the above assumptions only one of the Fermi surfaces would have a significantly field dependent effective mass. In this case the resistivity of only one Fermi surface would show a peak as a function of magnetic field. A commonly raised concern is that this peak should in a simple model not dominate the overall resistance since the conductance channels of all Fermi surface sheets add in parallel. This is seemingly at odds with the results by Grigera et al. [34], where a significant increase of at least a factor of 5 was observed in the A coefficient of the temperature dependent part of the resistivity as a function of magnetic field. However, one has to keep in mind that Grigera et al. have measured the A coefficient of the *magneto*-resistance of $Sr_3Ru_2O_7$. Therefore, one has to take into account that it is not the scalar longitudinal magnetoresistance of the Fermi surfaces but the full resistivity tensor that is added in parallel in order to calculate the overall resistivity of the material as discussed in Sect. 2.1.3.[7] Under these conditions it is possible that the divergence in the resistivity in one of the Fermi surfaces is dominating the total resistivity over a wide field range.

2.3 Summary

In this chapter I discussed the general framework of Fermi liquid theory as well as the concepts of quantum criticality and 'electron-nematic' transport. I furthermore reviewed the experimental evidence for a quantum critical end point in the material as well as thermodynamic measurements of a novel quantum phase in its vicinity which shows features of electron nematic like properties in transport measurements. I finally discussed the current knowledge of the zero field Fermi surface and the extent to which measured quasiparticle observations can account for the observed thermodynamic behaviour.

In this thesis several open questions arising from the current available experimental data will be addressed by magnetothermal studies and specific heat measurements.

---

[7] It has to be noted here that the simple addition of the resistivity tensors only holds for closed Fermi surfaces. If open orbits are present, the longitudinal magnetoresistance does not only depend on the magnitude of the applied magnetic field but also crucially on the orientation between the electric field and the open orbit, leading to significant anisotropies in transport. Intriguingly, if domains of different orbital orientations exist in the anomalous phase region that are orientable by a small in-plane magnetic field one would naturally expect 'nematik-like' transport properties similar to the ones observed by Borzi et al. To the authors best knowledge though discussed in private communications this path has not been explored in the theoretical literature so far.

The first part of the project is to investigate the entropic properties of the supposed low and high field Fermi liquid states. In particular the question of a peak in the Sommerfeld coefficient that was indirectly inferred from the measured peak in the A coefficient will be discussed. Furthermore data will be presented on magnetothermal oscillations, quantum oscillations in the entropy which can be studied by a method that is particularly sensitive to low frequencies.

The second part of the project relates to the novel quantum phase. Here the main data that will be presented will concern the nature of the surrounding phase transitions as well as the temperature dependence of the specific heat inside the phase.

References

1. Landau LD (1956) The theory of a Fermi liquid. JETP 30:1058
2. Landau LD (1957) Oscillations in a Fermi liquid. JETP 32:59
3. Landau LD (1958) On the theory of a Fermi liquid. JETP 35:97
4. Bednorz JG, Müller KA (1986) Possible high-T_c superconductivity in the Ba–La–Cu–O system. Z Phys B Condens Matter 64(2):189–193
5. Kamihara Y, Watanabe T, Hirano M, Hosono H (2008) Iron-based layered superconductor La[$O_{1-x}F_x$]FeAs ($x = 0.05 - 0.12$) with $T_c = 26$ K. J Am Chem Soc 130(11):3296–3297
6. Ashcroft NW, Mermin ND (1976) Solid state physics. Saunders, Philadelphia
7. Abrikosov AA, Gorkov LP, Dzyaloshinski IE (1975) Methods of quantum field theory in statistical physics. Dover, New York
8. The fermi surface data base (2009). http://www.phys.ufl.edu/fermisurface/jpg/Cu.jpg
9. Choy T-S, Naset J, Chen J, Hershfield S, Stanton C (2000) A database of fermi surface in virtual reality modeling language (vrml). Bulletin of the American Physical Society, vol 45, p L36.42
10. Shoenberg D (1984) Magnetic oscillations in metals. Cambridge University Press, Cambridge
11. Onsager L (1952) Philos Mag 43:1006
12. Pippard AB (1965) The dynamics of conduction electrons. Blackie, Glasgow
13. Kadowaki K, Woods SB (1986) Universal relationship of the resistivity and specific-heat in heavy-Fermion compounds. Solid State Commun 58(8):507–509
14. Hussey NE (2005) Non-generality of the Kadowaki–Woods ratio in correlated oxides. J Phys Soc Jpn 74(4):1107–1110
15. Landau LD (1960) Theoretical physics in the twentieth century, a memorial volume to Wolfgang Pauli. Interscience, NY, 245 pp
16. Kohn W, Luttinger JM (1965) New mechanism for superconductivity. Phys Rev Lett 15:524
17. Doiron-Leyraud N, Proust C, LeBoeuf D, Levallois J, Bonnemaison J-B, Liang R, Bonn DA, Hardy WN, Taillefer L (2007) Quantum oscillations and the Fermi surface in an underdoped high-T_c superconductor. Nature 447(7144):565–568
18. Sebastian SE, Harrison N, Palm E, Murphy TP, Mielke CH, Liang R, Bonn DA, Hardy WN, Lonzarich GG (2008) A multi-component Fermi surface in the vortex state of an underdoped high-T_c superconductor. Nature 454(7201):200–203
19. Vignolle B, Carrington A, Cooper RA, French MMJ, Mackenzie AP, Jaudet C, Vignolles D, Proust C, Hussey NE (2008) Quantum oscillations in an overdoped high-T_c superconductor. Nature 455(7215):952–955
20. Tsui DC, Störmer HL, Gossard AC (1982) Two-dimensional magnetotransport in the extreme quantum limit. Phys Rev Lett 48(22):1559–1562

21. Stewart GR (2001) Non-Fermi-liquid behavior in d- and f-electron metals. Rev Mod Phys 73(4):797–855
22. Löhneysen Hv, Rosch A, Vojta M, Wölfle P (2007) Fermi-liquid instabilities at magnetic quantum phase transitions. Rev Mod Phys 79(3):1015–1075
23. Gegenwart P, Si Q, Steglich F (2008) Quantum criticality in heavy-fermion metals. Nat Phys 4(3):186–197
24. Grigera SA, Mackenzie AP, Schofield AJ, Julian SR, Lonzarich GG (2002) A metamagnetic quantum critical endpoint in $Sr_3Ru_2O_7$. Int J Modern Phys B 16(20–22):3258–3264 (4th conference on physical phenomena at high magnetic fields, Santa FE, New Mexico, 19–25 October 2001)
25. Custers J, Gegenwart P, Wilhelm H, Neumaier K, Tokiwa Y, Trovarelli O, Geibel C, Steglich F, Pepin C, Coleman P (2003) The break-up of heavy electrons at a quantum critical point. Nature 424(6948):524–527
26. Mathur ND, Grosche FM, Julian SR, Walker IR, Freye DM, Haselwimmer RKW, Lonzarich GG (1998) Magnetically mediated superconductivity in heavy Fermion compounds. Nature 394(6688):39–43
27. Hertz JA (1976) Quantum critical phenomena. Phys Rev B 14(3):1165–1184
28. Millis AJ (1993) Effect of a nonzero temperature on quantum critical-points in itinerant Fermion systems. Phys Rev B 48(10):7183–7196
29. Maeno Y, Hashimoto H, Yoshida K, Nishizaki S, Fujita T, Bednorz JG, Lichtenberg F (1994) Superconductivity in a layered perovskite without copper. Nature 372(6506):532–534
30. Rice TM, Sigrist M (1995) Sr_2RuO_4 - an electronic analog of ^3He. J Phys Condens Matter 7(47):L643–L648
31. Mackenzie AP, Maeno Y (2003) The superconductivity of Sr_2RuO_4 and the physics of spin-triplet pairing. Rev Mod Phys 75(2):657–712
32. Ikeda S, Maeno Y, Nakatsuji S, Kosaka M, Uwatoko Y (2000) Ground state in $Sr_3Ru_2O_7$: Fermi liquid close to a ferromagnetic instability. Phys Rev B 62(10):R6089–R6092
33. Vollhardt D (1984) Normal ^3He—an almost localized Fermi-liquid. Rev Mod Phys 56(1):99–124
34. Grigera SA, Perry RS, Schofield AJ, Chiao M, Julian SR, Lonzarich GG, Ikeda SI, Maeno Y, Millis AJ, Mackenzie AP (2001) Magnetic field-tuned quantum criticality in the metallic ruthenate $Sr_3Ru_2O_7$. Science 294(5541):329–332
35. Grigera SA, Gegenwart P, Borzi RA, Weickert F, Schofield AJ, Perry RS, Tayama T, Sakakibara T, Maeno Y, Green AG, Mackenzie AP (2004) Disorder-sensitive phase formation linked to metamagnetic quantum criticality. Science 306(5699):1154–1157
36. Borzi RA, Grigera SA, Farrell J, Perry RS, Lister SJS, Lee SL, Tennant DA, Maeno Y, Mackenzie AP (2007) Formation of a nematic fluid at high fields in $Sr_3Ru_2O_7$. Science 315(5809):214–217
37. Mercure J-F (2008) The de Haas–van Alphen effect near a quantum critical end point in $Sr_3Ru_2O_7$. PhD Thesis, September 2008
38. Chmaissem O, Jorgensen JD, Shaked H, Ikeda S, Maeno Y (1998) Thermal expansion and compressibility of Sr_2RuO_4. Phys Rev B 57(9):5067–5070
39. Huang Q, Lynn JW, Erwin RW, Jarupatrakorn J, Cava RJ (1998) Oxygen displacements and search for magnetic order in $Sr_3Ru_2O_7$. Phys Rev B 58(13):8515–8521
40. Kiyanagi R, Tsuda K, Aso N, Kimura H, Noda Y, Yoshida Y, Ikeda S-I, Uwatoko Y (2004) Investigation of the structure of single crystal $Sr_3Ru_2O_7$ by neutron and convergent beam electron diffractions. J Phys Soc Jpn 73(3):639–642
41. Cao G, McCall S, Crow JE (1997) Observation of itinerant ferromagnetism in layered $Sr_3Ru_2O_7$ single crystals. Phys Rev B 55(2):R672–R675
42. Ikeda SI, Maeno Y (1999) Magnetic properties of bilayered $Sr_3Ru_2O_7$. Phys B 261:947–948 (international conference on strongly correlated electron systems (SCES 98), Paris, France, 15–18 July 1998)
43. Perry RS, Maeno Y (2004) Systematic approach to the growth of high-quality single crystals of $Sr_3Ru_2O_7$. J Cryst Growth 271(1–2):134–141

44. Perry RS, Galvin LM, Grigera SA, Capogna L, Schofield AJ, Mackenzie AP, Chiao M, Julian SR, Ikeda SI, Nakatsuji S, Maeno Y, Pfleiderer C (2001) Metamagnetism and critical fluctuations in high quality single crystals of the bilayer ruthenate $Sr_3Ru_2O_7$. Phys Rev Lett 86(12):2661–2664
45. Stryjewski E, Giordano N (1977) Metamagnetism. Adv Phys 26:487
46. Grigera SA, Borzi RA, Mackenzie AP, Julian SR, Perry RS, Maeno Y (2003) Angular dependence of the magnetic susceptibility in the itinerant metamagnet $Sr_3Ru_2O_7$. Phys Rev B 67(21):214427
47. Millis AJ, Schofield AJ, Lonzarich GG, Grigera SA (2010) Metamagnetic quantum criticality in metals. Phys Rev Lett 88(21):217204
48. Zhu LJ, Garst M, Rosch A, Si QM (2003) Universally diverging Grüneisen parameter and the magnetocaloric effect close to quantum critical points. Phys Rev Lett 91(6)
49. Berridge AM, Green AG, Grigera SA, Simons BD (2008) Inhomogeneous magnetic phases: a LOFF-like phase in $Sr_3Ru_2O_7$. Phys Rev Lett (accepted). arXiv:0810.2096v1
50. Gegenwart P, Weickert F, Garst M, Perry RS, Maeno Y (2006) Metamagnetic quantum criticality in $Sr_3Ru_2O_7$ studied by thermal expansion. Phys Rev Lett 96(13):136402
51. Nishizaki S, Maeno Y, Mao ZQ (2000) Changes in the superconducting state of Sr_2RuO_4 under magnetic fields probed by specific heat. J Phys Soc Jpn 69(2):572–578
52. Mackenzie AP, Ikeda S, Maeno Y, Fujita T, Julian SR, Lonzarich GG (1998) Fermi surface topography of Sr_2RuO_4. J Phys Soc Jpn 67(2):385–388
53. Bergemann C, Julian SR, Mackenzie AP, NishiZaki S, Maeno Y (2000) Detailed topography of the Fermi surface of Sr_2RuO_4. Phys Rev Lett 84(12):2662–2665
54. Mackenzie AP, Julian SR, Diver AJ, McMullan GJ, Ray MP, Lonzarich GG, Maeno Y, Nishizaki S, Fujita T (1996) Quantum oscillations in the layered perovskite superconductor Sr_2RuO_4. Phys Rev Lett 76(20):3786–3789
55. Bergemann C, Mackenzie AP, Julian SR, Forsythe D, Ohmichi E (2003) Quasi-two-dimensional Fermi liquid properties of the unconventional superconductor Sr_2RuO_4. Adv Phys 52(7):639–725
56. Damascelli A, Lu DH, Shen KM, Armitage NP, Ronning F, Feng DL, Kim C, Shen ZX, Kimura T, Tokura Y, Mao ZQ, Maeno Y (2000) Fermi surface, surface states, and surface reconstruction in Sr_2RuO_4. Phys Rev Lett 85(24):5194–5197
57. Mazin II, Singh DJ (1997) Ferromagnetic spin fluctuation induced superconductivity in Sr_2RuO_4. Phys Rev Lett 79(4):733–736
58. Shen KM, Kikugawa N, Bergemann C, Balicas L, Baumberger F, Meevasana W, Ingle NJC, Maeno Y, Shen Z-X, Mackenzie AP (2007) Evolution of the Fermi surface and quasiparticle renormalization through a van Hove singularity in $Sr_{2-y}La_yRuO_4$. Phys Rev Lett 99(18):187001
59. Kikugawa N, Mackenzie AP, Bergemann C, Borzi RA, Grigera SA, Maeno Y (2004) Rigid-band shift of the Fermi level in the strongly correlated metal: $Sr_{2-y}La_yRuO_4$. Phys Rev B 70(6):060508
60. Luttinger JM (1960) Fermi surface and some simple equilibrium properties of a system of interacting Fermions. Phys Rev 119(4):1153–1163
61. Tamai A, Allan MP, Mercure JF, Meevasana W, Dunkel R, Lu DH, Perry RS, Mackenzie AP, Singh DJ, Shen Z-X, Baumberger F (2008) Fermi surface and van Hove singularities in the itinerant metamagnet $Sr_3Ru_2O_7$. Phys Rev Lett 101(2):026407
62. Perry RS, Kitagawa K, Grigera SA, Borzi RA, Mackenzie AP, Ishida K, Maeno Y (2004) Multiple first-order metamagnetic transitions and quantum oscillations in ultrapure $Sr_3Ru_2O_7$. Phys Rev Lett 92(16):166602
63. Borzi RA, Grigera SA, Perry RS, Kikugawa N, Kitagawa K, Maeno Y, Mackenzie AP (2004) de Haas–van Alphen effect across the metamagnetic transition in $Sr_3Ru_2O_7$. Phys Rev Lett 92(21):216403
64. Damascelli A, Hussain Z, Shen ZX (2003) Angle-resolved photoemission studies of the cuprate superconductors. Rev Mod Phys 75(2):473–541

Chapter 3
Thermodynamic Measurements of Entropy

"My greatest concern was what to call it. I thought of calling it 'information', but the word was overly used, so I decided to call it 'uncertainty'. When I discussed it with John von Neumann, he had a better idea. Von Neumann told me, 'You should call it entropy, for two reasons. In the first place your uncertainty function has been used in statistical mechanics under that name, so it already has a name. In the second place, and more important, nobody knows what entropy really is, so in a debate you will always have the advantage.'"—Claude Shannon (1948) [1].

In the experiments presented in this thesis the magnetic phase diagram of $Sr_3Ru_2O_7$ is studied in detail. Of particular interest are the changes of the entropy of this system with both temperature and magnetic field. Entropy is certainly one of the less tangible concepts of physics, its understanding furthermore complicated by the variety of ways in which it can be defined. This however should not lead one to underestimate its central importance to thermodynamic studies. Not only is it fundamental to the definition of a temperature scale in thermodynamics. It also is one if not *the* quantity linking the field of thermodynamics with the subject of statistical physics via the postulate that the entropy S of a system with N accessible microstates is given in the microcanonical ensemble by

$$S = k_B \ln N. \tag{3.1}$$

Here k_B is as usually the Boltzmann constant. Entropy is therefore the quantity that links the theoretical predictions of microscopic models with the experimentally accessible quantities of thermodynamics.

Furthermore, entropy and its derivatives such as specific heat play an important role in the study of phase transitions. As will be discussed in the following sections entropy usually exhibits a step change across first order transitions leading to the absorption or release of heat—the so called latent heat. Across second order transitions on the other hand entropy evolves continuously but not smoothly, leading to discontinuities or divergences in the derivatives of the free energy such as the specific heat. Measurements of specific heat are therefore fundamental to the study of phase transitions.

In the following I will introduce the necessary theoretical background to thermodynamic measurements under magnetic fields. This will be followed by a description of the measurement methods for the specific heat and magnetocaloric effect used in the experiments discussed in Chaps 4 and 5.

3.1 General Considerations on Thermodynamics in Magnetic Fields

In this section the main thermodynamic concepts relevant to this work will be discussed. The intention is not to give an introduction on thermodynamics, which is well covered in a variety of textbooks. However, there exist differences across the literature regarding the exact definition of such central quantities as magnetic work. Therefore, in the following sections an overview is given of some of the most important aspects of thermodynamics in magnetic fields.

3.1.1 The Laws of Thermodynamics for Magnetic Systems

It is assumed that the reader is familiar with the general aspects of thermodynamics. Here, the laws of thermodynamics will be briefly discussed regarding their specific formulation for magnetic systems and relevant implications for this work.

3.1.1.1 The First Law: Energy Conservation

The first law of thermodynamics is in principle a statement about the conservation of energy. In its mathematical form it is expressed as

$$dU = \dbar W + \dbar Q, \tag{3.2}$$

where dU is the change of internal energy during an infinitesimal change of state during which the amount of work $\dbar W$ is done to and the amount of heat $\dbar Q$ is absorbed by the system under consideration.

For magnetic systems it is in particular the magnetic work term $\dbar W$ for which differing definitions are given in textbooks. The fundamental question for the experiments presented in this work is what work $\dbar W$ has to be done to magnetise a sample inside a coil. A very good discussion of this is given for

3.1 General Considerations on Thermodynamics in Magnetic Fields

example in the book by Mandl [2]. There it is shown that the total work done is given by[1]

$$đW = \mu_0 H dM. \tag{3.3}$$

Here μ_0 is the magnetic vacuum permeability. The above definition includes the mutual interaction energy $\mu_0 HM$ between the vacuum field H and the magnetisation M. It is sometimes argued that this is not part of the *physical* system represented by the sample alone and should be excluded. That indeed simplifies many problems from a conceptual and calculational point of view in statistical mechanics. However $\mu_0 HM$ is an energy that has to be changed if one wants to change the thermodynamic state of the sample. It is therefore to be included if one considers the *thermodynamic* system onto which work is done, so that is the definition that will be used in this thesis.

For a purely magnetic system the first law of thermodynamics therefore reads

$$dU = đQ + \mu_0 H dM. \tag{3.4}$$

3.1.1.2 The Second Law

> Any method involving the notion of entropy, the very existence of which depends on the second law of thermodynamics, will doubtless seem to many far-fetched, and may repel beginners as obscure and difficult of comprehension.—Attributed to Gibbs (1873) [3].

The second law introduces entropy as a property of the thermodynamic state of a system. It is based on the observation that for any closed reversible thermodynamic path

$$\oint \frac{đQ}{T} = 0 \tag{3.5}$$

holds. Therefore, one can define a new variable S that is a thermodynamic function of state of the system and whose differential is given by $dS = đQ/T$.

Furthermore, let us assume that a given system in an experiment has as its natural variables temperature T and externally applied magnetic field H. Then the entropy is a function of T and H only.[2] The total differential of the entropy can therefore be written as

---

[1] This expression excludes the vacuum self energy term, which is independent of the thermodynamic state of the system under consideration.
[2] Here I ignore $-PdV$ terms since all experiments take place at $P = 0$ in vacuum.

$$dS(T,H) = \left.\frac{\partial S}{\partial T}\right|_H dT + \left.\frac{\partial S}{\partial H}\right|_T dH, \tag{3.6}$$

where

$$\left.\frac{\partial S}{\partial T}\right|_H = \left.\frac{C_H}{T}\right|_H \tag{3.7}$$

is the heat capacity of the system at constant magnetic field, C_H, divided by temperature T. This quantity describes how much the heat content of a system changes upon a small change of temperature. Similarly, one can measure the change of entropy while changing the magnetic field at constant temperature. This leads to the magnetocaloric coefficient

$$\Gamma = T\left.\frac{\partial S}{\partial H}\right|_T. \tag{3.8}$$

Both quantities, the specific heat and the magnetocaloric coefficient, enclose similar important information—how the entropy of a system changes as a function of temperature or applied magnetic field.

Finally it should be mentioned that the first and second law of thermodynamics allow us to write down the magnetic version of one of Maxwell's relations

$$\left.\frac{1}{\mu_0}\frac{\partial S}{\partial H}\right|_T = \left.\frac{\partial M}{\partial T}\right|_H. \tag{3.9}$$

3.1.1.3 Consequences of The Third Law

The third law of thermodynamics is the most subtle (and indeed controversial) one. Originally derived in 1906 by Nernst, it is an experimental observation stating that at zero temperature all thermodynamic systems have the same entropy, to be defined as zero. From a theoretical point of view this law has a purely quantum mechanical origin and was contested initially.

Since at zero temperature all entropies are zero, it has in particular to hold that

$$\lim_{T\to 0}\frac{\partial S}{\partial H} = 0, \tag{3.10}$$

and through the Maxwell relation (3.9)

$$\lim_{T\to 0}\frac{\partial M}{\partial T} = 0. \tag{3.11}$$

3.1.2 Phase Transitions

The previous section introduced the basic relations for thermodynamic states of magnetic systems. A further important aspect of thermodynamic experiments is the study of phase transitions between these states.

First and Second Order Transitions
The physics of phase transitions is a very well-studied subject and good introductions can be found in several textbooks. This section will concentrate in particular on the physics of first order transitions in magnetic fields as this is relevant for the analysis in Chap. 5.

Clausius–Clapeyron Relation
The well-known Clausius–Clapeyron relation for a PVT system relates the curvature of a first order phase transition line in the P–T diagram, dP_C/dT_C, with the difference in entropy ($S_2 - S_1$) and volume ($V_2 - V_1$) between the two states 1 and 2 coexisting at the pressure P_C and temperature T_C. The relation is given by

$$\frac{dP_C}{dT_C} = \frac{(S_2 - S_1)}{(V_2 - V_1)}. \tag{3.12}$$

Since it is not commonly found in textbooks, the derivation of this relation for the equivalent magnetic transition will be presented in the following.

The starting point for the derivation is the Gibbs free energy G, which is the appropriate thermodynamic potential if the natural variables are magnetic field H and temperature T such as in the experiments discussed here. It can be expressed in terms of the internal energy U, temperature T, entropy S, magnetic field H and magnetisation M as

$$G = U - TS - \mu_0 MH = \mu N, \tag{3.13}$$

with μ being the chemical potential and N the number of constituting particles. At a first order phase transition two phases (1) and (2) are in thermal equilibrium and $\mu_1 = \mu_2 = \mu$ has to hold. Therefore, the Gibbs free energies G have to be equal for both phases and the line of first order transitions is given by $G_1(H, T) = G_2(H, T)$.[3] From the definition of magnetic work the form of the

---

[3] The general principle is, that at the transition the appropriate thermodynamic potential, whose variables are the natural variables of the experimental conditions, has to be identical for the two phases. In the case here the natural variables are magnetic field and temperature, which requires the Gibbs free energies of the two phases to be equal.

differential of the internal energy, dU, was derived in Eq. 3.4. It therefore follows that the differential of the Gibbs free energy for a magnetic system is given by

$$dG_{1,2} = -S_{1,2}dT - \mu_0 M_{1,2}dH + \mu dN. \tag{3.14}$$

Along the first order transition line both G_1 and G_2 have to change by equal amounts since along the transition line $G_1 = G_2$ holds. It follows therefore together with Eq. 3.14 that

$$-S_1 dT_C - \mu_0 M_1 dH_C = -S_2 dT_C - \mu_0 M_2 dH_C \tag{3.15}$$

is true at the transition. Here the index c indicates that the equation only holds for the critical values of the temperature and magnetic field at which the transition occurs. This equation can now be rewritten to give the magnetic form of the Clausius–Clapeyron relation:

$$\mu_0 \frac{dH_C}{dT_C} = -\frac{(S_2 - S_1)}{(M_2 - M_1)}. \tag{3.16}$$

Here $\mu_0 dH_C/dT_C$ gives the curvature of the first order transition line in the H–T plane of the phase diagram. One direct consequence concerns the curvature of the phase transition line at zero temperature. In combination with the third law of thermodynamics it follows that ΔS is zero at absolute zero. In a first order metamagnetic transition ΔM is in general not zero. Therefore, the curvature dH_C/dT_C has to be zero.

3.2 Experimental Consequences

The basis for measuring changes in the entropy of a system used in this thesis is the thermodynamic relation

$$dS = \frac{đQ}{T}. \tag{3.17}$$

Here $đQ$ is a heat flux in or out of the sample at temperature T resulting in a differential change of entropy dS in the sample. This simple equation is at the heart of thermodynamics and allows us to extract changes in the entropy during an experiment.

In Fig. 3.1 the schematic setup of an experiment based on Eq. (3.17) is shown. One has to be able to apply heat $đQ$ to the system in a controlled way via a heater and to measure the temperature of the sample via a thermometer. For the purposes of the following discussion it will be assumed that the specific heat of the thermometer and the heater are zero and the thermal link between them and the sample is infinitely good. The temperature of the bath is supposed to be constant. The system is therefore characterised by the specific heat of the sample and the thermal conductance of the link.

3.2 Experimental Consequences

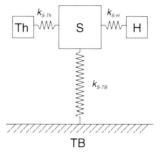

Fig. 3.1 Thermal schematic of a typical caloric experiment showing its main components and their relation to each other. The temperature of the sample *S* is measured by the thermometer *Th*. A controlled amount of heat dQ is applied to the sample via the heater *H*. The thermal bath *TB* provides a heat reservoir at fixed temperature T_0. The components are linked via thermal links of conductances k_{S-Th}, k_{S-H} and k_{S-TB} as shown

3.2.1 Principle of Specific Heat Measurements

Heat as a form of energy and the measurement of specific heat of different substances have been at the core of thermodynamics. The original principle of measuring the specific heat by investigating the change of temperature after supplying a well-defined amount of energy to the substance under investigation is still used in so called adiabatic measurements. However, since the 1960s other non-adiabatic methods have been developed, in which the sample is not completely isolated. In the following the three methods employed in this work will be introduced. These are the main methods available and different experiments are often only different in the explicit realisation of the generic components, i.e. the heater, thermometer and thermal links. The interested reader can find more detailed information on specific heat measurements for example in the review by Stewart [4].

3.2.1.1 Adiabatic Measurement

The adiabatic method goes back to the definition of the heat capacity of a thermodynamic system as

$$C_{\text{System}}(T) = \lim_{\Delta T \to 0} \left(\frac{\Delta Q_{\text{input}}}{\Delta T} \right) \tag{3.18}$$

with ΔQ_{input} being a heat pulse applied to the sample at a time t_0 and ΔT being the temperature difference between the initial temperature T_0 and the temperature T_1 to which the sample is heated due to the heat pulse.

Therefore, by applying a small amount of heat ΔQ and measuring the temperature change ΔT one can calculate the specific heat. The characterisation of this method as 'adiabatic' already implies the main technological challenge: the experimental setup has to be thermally decoupled from the bath in order to avoid exchange of heat between the sample and the bath during the experiment.

Temperature differences between the setup and the environment cause constant heat flows which distort the temperature measurement. Therefore the thermal conduction k_{S-TB} has to be chosen to be small in order to minimize the heat transfer. This applies as well to the additional heat leaks caused by connecting wires for heater and thermometer to the experimental environment. The experimental reality is shown in Fig. 3.2. Here the sample temperature T is plotted in blue as a function of time for an idealised experiment. Initially the sample is at temperature T_0. At time t_0 a heat pulse is applied. This causes the temperature of the sample to rise to T_1. After the jump in temperature the sample decays exponentially back to its equilibrium temperature T_0.

In a realistic experiment the jump has a finite width in time as indicated in red. Its height can be extracted using the so-called midpoint construction by fitting an exponential curve to the later times $t > t_0$.

However, especially at very low temperatures (below 1 K) it becomes increasingly difficult to sufficiently thermally isolate the sample and at the same time to be able to cool it down within a reasonable amount of time. It becomes essentially impossible to measure more than a few data points within reasonable times. A major breakthrough came around 1970 with two important papers by Sullivan and Seidel [5] as well as Bachmann et al. [6] experimentally realising specific heat measurements that did not require a near perfect thermal isolation from the environment. These methods are called 'non-adiabatic' for that reason and are introduced in the following sections.

3.2.1.2 Relaxation Time Method

The most direct extension of the adiabatic method was first introduced by Bachmann et al. [6]. They employed a measurement procedure schematically shown in Fig. 3.3. Here the temperature of the sample is plotted as a function of time during a measurement cycle. Initially (phase I) the sample is in equilibrium with the thermal bath at temperature T_0. At time t_1 a continuous heating is applied and sustained during the whole of phase II. This causes the sample to heat up and reach

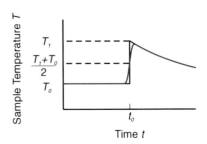

Fig. 3.2 Sample temperature as a function of time during a heat pulse experiment. In *blue* the idealised experiment with instantaneous heat jump is shown. In *red* a more realistic curve in an experiment with a finite width in time for the temperature jump is given

3.2 Experimental Consequences

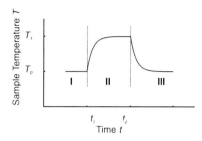

Fig. 3.3 Trace of the sample temperature T as a function of time t during a relaxation time measurement cycle. In phases *I* and *III* no heat is applied whereas during phase *II* the sample is heated by a continuous heat input

a new temperature T_1. When this new steady state is reached, within experimental resolution, the heating is switched off at time t_2 and the sample allowed to relax back to equilibrium with the bath.

It was shown by Bachmann et al. [6] that the following two relations can be derived by solving the differential equations describing the temperature dependence of the sample.

First, the temperature rise at the beginning of phase II is exponential in nature. The temperature T of the sample as a function of time is given by

$$T(t) = T_1 + (T_0 - T_1)e^{-\frac{k}{C}(t-t_1)} \quad \text{for } t_1 < t < t_2, \quad (3.19)$$

with C being the heat capacity of the sample. Equivalently, it has to hold for the temperature decay of phase III that

$$T(t) = T_0 + (T_1 - T_0)e^{-\frac{k}{C}(t-t_2)} \quad \text{for } t > t_2. \quad (3.20)$$

Secondly, the thermal link k of the system is given independently through the applied power $\dot{Q}$ by

$$k = \frac{\dot{Q}}{T_1 - T_0}. \quad (3.21)$$

Together, these two relations allow the measurement of the heat capacity C by extracting the time constant $\tau = C/k$ from the exponential relaxation of the sample together with the independent measurement of the thermal link k. A modification of the above method [7] is the primary technique used by the Quantum Design PPMS system.

3.2.1.3 AC Method

A major disadvantage of the above relaxation time method is that it is still a 'discrete' measurement at a fixed temperature of the thermal bath. It is therefore not very well suited for measuring significant changes in specific heat over small changes of control parameter such as temperature or magnetic field—a situation usually found at phase transitions. The final method presented here is a breakthrough in improvement of the relative accuracy of specific heat measurements and was pioneered by Sullivan and Seidel [5].

The most important change introduced by them was to apply an oscillating heater power to the sample in the form of

$$\dot{Q} = \dot{Q}_0 \cos(\omega t), \tag{3.22}$$

with $\dot{Q}_0$ the amplitude of the oscillation, $\omega = 2\pi f$ the angular frequency and t the time. This can for example be simply achieved by using a resistive heater and applying an AC current of frequency $f/2$.

For the analysis of the experiment a slightly more detailed description of the setup than previously used is necessary. In Fig. 3.4a the basic geometry used in the experiments reported in this thesis is shown. The power $\dot{Q}$ is applied to one side of a sample of length L and cross section A. The temperature is measured on the other side, connected via a thermal link k to the thermal bath.

There are two important time scales in this system. The first is τ_{int} which is the characteristic time scale for heat diffusion inside the sample. The second is τ_2 which is the thermal time constant between the sample and the thermal bath. It is given by the coefficient of the heat capacity of the sample C and the thermal

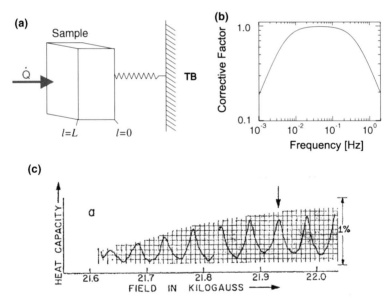

Fig. 3.4 **a** The geometry for the AC specific heat measurements performed in this project. The oscillating heat input $\dot{Q}$ is applied to one side of a sample of thickness L. The temperature is measured on the other side at $l = 0$. This side is also weakly coupled to the thermal bath *TB*. **b** The factor describing the corrections to the amplitude in Eq. 3.23 as a function of frequency $f = \omega/2\pi$ due to the finite time constants τ_{int} and τ_2 of the setup. **c** An example of the possible relative accuracy with this method [5]. Here, quantum oscillations in the heat capacity of a Beryllium sample were resolved as a function of magnetic field

3.2 Experimental Consequences

conductance of the link k, i.e. $\tau_2 = C/k$. From the way the experiment is set up it follows that one tries to achieve $\tau_{int} \ll \tau_2$ so that the sample reaches internal equilibrium on much shorter time scales than with the thermal bath.

If the frequency ω of the applied AC heating satisfies the condition $\omega\tau_{int} \ll 1 \ll \omega\tau_2$ one can show that the temperature of the sample is oscillating around an average temperature $T_{average} > T_0$ with the oscillatory component $\tilde{T}$ given by[4]

$$|\tilde{T}| \approx \frac{\dot{Q}}{2\omega C}\left(1 + \omega^2 \tau_{int}^2 + \frac{1}{\omega^2\tau_2^2} + \frac{2k}{3k_S}\right)^{1/2}. \tag{3.23}$$

Here $\dot{Q}^2$ is the amplitude of the heat input oscillating at angular frequency ω and C the heat capacity of the sample. Furthermore, k_S is the thermal link represented by the sample itself.

The terms in the bracket describe several corrections to the amplitude of the oscillations. The second term, containing the internal relaxation time τ_{int}, describes the thermal damping along the length of the sample. The higher the applied frequency ω is, the more effective this damping becomes. The third term, containing the time constant τ_2, corresponds to a correction to the size of the amplitude of the oscillations due to the fact that heat can flow in and out of the system through the thermal link. Finally, the third term is a frequency independent correction depending on the ratio k/k_S, where k is the thermal link between the sample and bath and k_S the thermal link represented by the sample itself. It is included here for completeness but is negligible in the experimental setup discussed later in this work because of $k_S \gg k$.

In Fig. 3.4b the resulting typical frequency dependence of the corrective factor is shown.[5] In an experiment the optimal frequency is that of the maximum of this curve, since here the corrections to the signal are minimal. This furthermore ensures that changes in specific heat have a negligible impact on the overall corrective terms which depend indirectly on the heat capacity of the sample.

Therefore, in the above described limit the amplitude of temperature oscillations is to first approximation directly proportional to the inverse of the heat capacity of the sample C. It is this fact that is used to measure C. By continuously monitoring the temperature oscillations while slowly changing a control parameter such as the thermal bath temperature or the magnetic field it is possible to measure the specific heat of the sample with high relative precision. An example of the achievable precision is given in Fig. 3.4c. Here the heat capacity C of a Beryllium sample is shown as a function of applied field [5]. The observable oscillations are quantum oscillations in C.

---

[4] Here it is assumed as before that the heater and thermometer have a negligible heat capacity and are in thermal equilibrium with the sample.

[5] Values for the time constants are chosen corresponding to the experiment described later.

3.2.2 Magnetocaloric Measurements

In the previous section, measurement methods for the specific heat c of a material were discussed in detail. As shown in Sect. 3.1.1 specific heat effectively gives information on how the entropy of a system changes with temperature. In the following, methods are discussed that measure the isothermal change of entropy while changing an applied magnetic field. In this case one is therefore interested in

$$\left.\frac{\partial S}{\partial H}\right|_T. \tag{3.24}$$

It is often argued that the isothermal change in the entropy S between two magnetic fields H_0 and H_1 can be calculated from measurements of the heat capacity C of a sample. Indeed, according to the third law of thermodynamics, at zero temperature all phases have, independent of magnetic field H, the same entropy $S_{T=0}(H) = 0$ as a function of magnetic field. Therefore, one can in principle write

$$S(H_1, T) - S(H_0, T) = \int_{T'=0}^{T'=T} \left\{ \frac{\partial S(H_1, T')}{\partial T'} - \frac{\partial S(H_0, T')}{\partial T'} \right\} dT', \tag{3.25}$$

and consequently

$$S(H_1, T) - S(H_0, T) = \int_{T'=0}^{T'=T} \left\{ \frac{C(H_1, T')}{T'} - \frac{C(H_0, T')}{T'} \right\} dT'. \tag{3.26}$$

This shows that theoretically it is possible to reconstruct the entropy of a system over the whole phase diagram by only measuring the specific heat $C(H, T)$ as a function of the appropriate magnetic fields and temperature. However, it is not possible to measure C to zero temperature. One therefore has to extrapolate high temperature measurements down to absolute zero in order to use the above equation. If the material is in a thermodynamically well understood state such as a Fermi liquid, this is still possible since the temperature dependence can be assumed to be known. However this method breaks down in particular close to magnetic phase transitions and in thermodynamic phases that are not well understood, i.e. in exactly the most interesting cases from a scientific point of view. Here the magnetocaloric effect becomes important since it can, as will be shown below, measure entropy changes as a function of field without requiring experimentally inaccessible information from lower temperatures.

In order to see how this can be extracted from an experimental setup such as that shown in Fig. 3.1 one has to analyse the heat flow occurring during a change of magnetic field at a constant rate in time.

3.2 Experimental Consequences

First of all heat dQ_{link} can flow in time dt from or to the sample through the thermal link between the sample and the thermal bath. dQ_{link} is given by

$$\frac{dQ_{link}}{dt} = -k(T_S - T_B), \quad (3.27)$$

with k being the thermal conductance of the heat link, T_S the sample temperature and T_B the temperature of the thermal bath. The sign convention here is such that dQ_{link} is negative when heat flows from the sample. According to the second law of thermodynamics it is equal to $T_S dS$, resulting in

$$T_S \frac{dS}{dt} = -k(T_S - T_B). \quad (3.28)$$

Furthermore, since the entropy S of the system is a function of temperature T and magnetic field H only, Eq. 3.28 can be rewritten as

$$T_S \left\{ \frac{\partial S}{\partial T}\bigg|_H \frac{dT_S}{dt} + \frac{\partial S}{\partial H}\bigg|_T \frac{dH}{dt} \right\} = -k(T_S - T_B). \quad (3.29)$$

If the magnetic field at the sample is changed by dH in a time interval dt, then one can express the total differential of the sample temperature with respect to time as

$$\frac{dT_S}{dt} = \frac{dT_S}{dH}\frac{dH}{dt}. \quad (3.30)$$

Applying this relation to Eq. 3.29 gives

$$\left\{ T_S \frac{\partial S}{\partial T}\bigg|_H \frac{dT}{dH} + T_S \frac{\partial S}{\partial H}\bigg|_T \right\} \frac{dH}{dt} = -k(T_S - T_B). \quad (3.31)$$

This can further be simplified by using the definition for the heat capacity C, which is

$$C = T_S \frac{\partial S}{\partial T}\bigg|_H, \quad (3.32)$$

resulting in

$$\left\{ C\frac{dT}{dH} + T_S \frac{\partial S}{\partial H}\bigg|_T \right\} \frac{dH}{dt} = -k(T_S - T_B). \quad (3.33)$$

The final step is to rearrange the above equation. This results in an expression for the partial differential of the entropy S with respect to the magnetic field H given by

$$\frac{\partial S}{\partial H}\bigg|_T = -\frac{k}{T_S}(T_S - T_B)\frac{1}{\dot{H}} - \frac{C}{T_S}\frac{dT_S}{dH}. \quad (3.34)$$

In the following I will consider two cases. The first is the adiabatic condition. Though not the one used for this experiment it is the one most usually encountered and should be discussed for completeness.

The second is the limit of a good thermal link between the sample and the thermal bath. This limit allows nearly isothermal measurements. It is slightly more complex to implement but overcomes some of the limitations of adiabatic measurements.

3.2.2.1 Adiabatic Conditions

Under adiabatic conditions the thermal link between the sample and the bath can be considered to be zero, i.e. no heat can be exchanged with the environment. Equation 3.34 therefore changes to

$$\left.\frac{\partial S(T_S, H)}{\partial H}\right|_{T_S} = -\frac{C}{T_S}\frac{dT_S}{dH}. \tag{3.35}$$

This can be rewritten as

$$dT_S = -\frac{T_S}{C}\left.\frac{\partial S(T_S, H)}{\partial H}\right|_{T_S} dH. \tag{3.36}$$

In other words a change of magnetic field by dH will cause a change in temperature dT_S of the sample that is proportional to $\partial S/\partial H$. This result is often also quoted in a similar form based on the Maxwell equation 3.9 given in Sect. 3.1.1

$$dT_S = -\mu_0 \frac{T_S}{C}\left.\frac{\partial M}{\partial T}\right|_H dH. \tag{3.37}$$

The two equations above describe the underlying physics in particular of adiabatic demagnetisation refrigeration. Here the temperature of a material such as a paramagnetic salt is changed by changing the applied magnetic field. This can then be used to cool an experimental setup. This well known principle is discussed in several textbooks such as [8] and [9].

The following discussion will concentrate on the use of the effect in studying the phase diagram of a material. One can, for example, consider the case of increasing the magnetic field H applied to a sample linearly as a function of time t, such that at $t = t_0$ the sample passes through a first order transition, as shown in Fig. 3.5a. At that moment a step change in entropy occurs and latent heat is released. In this case a step change in temperature will be observed as shown in Fig. 3.5b.

Therefore, this form of the magnetocaloric effect is well suited to track with high accuracy the phase diagram of a material by simply monitoring the temperature of the sample as a function of magnetic field during a field sweep. A very good example of this is shown in Fig. 3.6 [10].

3.2 Experimental Consequences

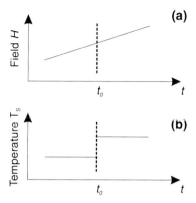

Fig. 3.5 Here the magnetocaloric effect is shown schematically for the case of crossing a first order transition line under adiabatic conditions. The magnetic field is increased linearly as a function of time t as indicated in (**a**). At time t_0 the sample undergoes a first order transition. This results in the release of latent heat associated with the first order transition causing the temperature of the sample to increase step like, as indicated in (**b**)

Fig. 3.6 Phase diagram of URu_2Si_2 in the $H-T$ plane [10]. The traces are the results of magnetocaloric sweeps in nearly adiabatic conditions with different initial sample temperatures

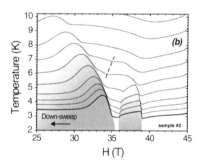

Here the temperature of the sample as a function of magnetic field during a field sweep is shown for several temperatures of the thermal bath. The discontinuities in the temperature traces at the phase boundaries are clearly identifiable.

However, it is less direct to extract further quantitative information about the entropy changes as a function of magnetic field due to the large temperature changes of the sample and the fact that true adiabatic conditions are difficult to achieve at low temperatures. The non-adiabatic method with a finite thermal link discussed in the following addresses these issues in particular.

3.2.2.2 Non-adiabatic Conditions

In Fig. 3.7 the same process as in the previous graph is shown. As indicated in part (a) the magnetic field at the position of the sample is increased as a function of time t such that at $t = t_0$ the sample undergoes a first order phase transition.

Fig. 3.7 The same experimental process as in Fig. 3.5 is shown for non-adiabatic conditions. Contrary to the previous case the sample temperature relaxes back to the equilibrium temperature after the magnetic field has been increased beyond the point of the first order transition

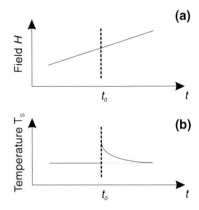

Part (b) shows the temperature of the sample as a function of time. Similarly to the previous case the temperature of the sample T_S undergoes a jump at t_0. However, contrary to the previous case the sample relaxes back to equilibrium with the thermal bath. Therefore this measurement has the same ability to trace the phase diagram as the previously discussed adiabatic method. It has the added advantage that the measurement is quasi-isothermal if the sample temperature stays sufficiently close to the thermal bath temperature.

So far the discussion has solely concentrated on the effect of phase transitions. However, changes in entropy can also occur inside a thermodynamic phase as a function of magnetic field. In Eq. 3.34 the connection between the temperature of the sample T_S and the change in entropy $\partial S/\partial H$ was derived to be

$$\left.\frac{\partial S}{\partial H}\right|_{T_S} = -\frac{k}{T_S}(T_S - T_B)\frac{1}{\dot H} - \frac{C}{T_S}\frac{dT_S}{dH}.$$

In the case where the sample is relatively well coupled to the thermal bath the second term can be neglected in the analysis compared to the first. Therefore, in the highly non-adiabatic case, one can write

$$\left.\frac{\partial S}{\partial H}\right|_{T_S} \approx -\frac{k}{T_S}(T_S - T_B)\frac{1}{\dot H}. \tag{3.38}$$

Since $\partial S/\partial H$ as well as k are independent of the direction in which the magnetic field is swept, whereas $1/\dot H$ is not, the term $(T_S - T_B)$ has to change sign. In other words the temperature trace of a sample during a magnetic field sweep with increasing field is opposite to the temperature trace during a field sweep of decreasing magnetic field.

This situation is depicted in Fig. 3.8a where a temperature trace of a $Sr_3Ru_2O_7$ sample is shown schematically during a sweep of the magnetic field with increasing field (blue) and decreasing field (red). Here the aim is not to give details of the physics of $Sr_3Ru_2O_7$ contained in these traces. This will be done in detail in Chap. 5. However, for the current discussion two things should be pointed out.

3.2 Experimental Consequences

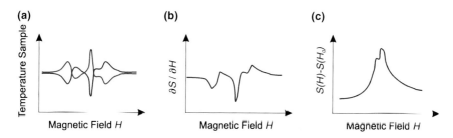

Fig. 3.8 **a** The temperature of the sample as a function of magnetic field during a sweep of the magnetic field for increasing fields (*blue*) and decreasing fields (*red*). Since one is in the well coupled limit the traces are seemingly mirror symmetric about the equilibrium temperature. **b** The reconstructed $\partial S/\partial H$ as a function of magnetic field based on the traces in **a**. **c** The entropy change as a function of magnetic field that results from integrating the trace in **b** with respect to a reference field H_0

First, the position of phase transitions is clearly visible in form of sharp features in the temperature traces. Second, the two traces appear to be reflections of each other about the average temperature. As discussed above, this is a direct consequence of the experiments being conducted in the highly non-adiabatic regime.

At the sharp transition features the temperature of the sample varies rapidly. In these parts of the traces the approximation of neglecting the second term in Eq. 3.34 breaks down. In consequence, in order to reconstruct the entropy of the sample along the whole trace, one has to analyse the full equation. In a first step $\partial S/\partial H$ is calculated based on the additional information of the thermal link k and the heat capacity of the sample C. The resulting $\partial S/\partial H$ as a function of magnetic field H is shown in Fig. 3.8b. This looks very similar to the temperature trace itself due to the fact that the first term that is directly proportional to $(T_S - T_b)$ dominates. The second step is then to integrate $\partial S/\partial H$ as a function of H with respect to a chosen reference field H_0. This way one obtains $S(H) - S(H_0)$ as a function of magnetic field H as shown in Fig. 3.8c.

If the rate of change of the magnetic field $\dot{H}$ during a sweep is chosen to be low enough then one can ensure that the sample temperature does not change more than a few percent around the average temperature. This is comparable to the changes in sample temperature in the previously presented specific heat measurement methods. One can therefore say within experimental possibilities that the reconstructed trace of the entropy change of the sample as a function of magnetic field is isothermal.

3.2.2.3 Magnetocaloric Oscillations

As has been discussed in (Sect. 2.1.2), Fermi liquids show quantum oscillations as a function of applied magnetic field in many observable quantities. Since the effect is created by an oscillatory component in the density of states at the Fermi energy it also has to be observable in the magnetocaloric effect. These oscillations have

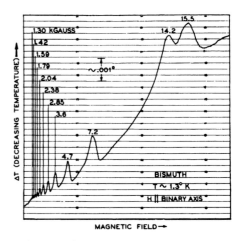

Fig. 3.9 Quantum oscillations in bismuth observed in the change of temperature during a magnetic field sweep [11]

indeed been measured and were found first, to my knowledge, by Boyle et al. [11] for bismuth. Their results are shown in Fig. 3.9.

The first theoretical analysis was published by Condon in his seminal paper from 1966 [12]. In the following, rather than deriving the effect from first principles it will be motivated based on the de Haas–van Alphen effect discussed in (Sect. 2.1.2).

As shown there, the total magnetization $M(H, T)$ of a Fermi liquid as a function of magnetic field H and temperature T can be expressed as the sum of a 'normal' non-oscillating background contribution M_{BG} and an oscillatory component $\tilde{M}$ whose origin is the phenomenon of quantum oscillations.

$$M(T, H) = M_{BG}(T, H) + \tilde{M}(T, H). \tag{3.39}$$

Furthermore, according to Maxwell's relation (see Eq. 3.9 given in Sect. 3.1.1) one knows that

$$\frac{1}{\mu_0} \left.\frac{\partial S}{\partial H}\right|_T = \left.\frac{\partial M}{\partial T}\right|_H. \tag{3.40}$$

Therefore, the entropy has to show an oscillatory component $\tilde{S}(H, T)$ that is related to the oscillations in magnetisation via

$$\frac{1}{\mu_0} \left.\frac{\partial \tilde{S}}{\partial H}\right|_T = \left.\frac{\partial \tilde{M}}{\partial T}\right|_H. \tag{3.41}$$

Furthermore, it was shown in the previous section that during a magnetocaloric sweep under non-adiabatic conditions, the change of entropy S with magnetic field H can be approximated as

$$\left.\frac{\partial \tilde{S}}{\partial H}\right|_T \approx -k(\tilde{T}_S - T_B)\frac{1}{H}.$$

3.2 Experimental Consequences

Fig. 3.10 Here the functional form of the Lifshitz–Kosevich formula (*blue*) and its temperature derivative (*red*) are shown as a function of temperature

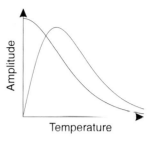

That this condition is fulfilled can be experimentally checked by the reflection symmetry between of the quantity $(T_S - T_B)$ between a sweep of increasing and decreasing magnetic field.

Therefore, for quantum oscillations in the magnetisation that have a sufficiently low frequency the relation

$$\mu_0 \frac{\partial \tilde{M}}{\partial T}\bigg|_H = -k(\tilde{T}_S - T_B)\frac{1}{H} \quad (3.42)$$

holds.

As shown in Sect. 2.1.2, the temperature dependence of the oscillations in magnetisation is given by the Lifshitz–Kosevich factor R_{LK}. Therefore, the amplitude of the magnetocaloric oscillations is proportional to the temperature derivative of the Lifshitz–Kosevich factor, i.e. $\partial R_{LK}/\partial T$ rather than R_{LK} itself. The function is given by

$$\frac{\partial R_{LK}}{\partial T} = \frac{\partial}{\partial T}\left\{\frac{x}{\sinh(x)}\right\} = \left\{\frac{1}{\sinh(x)} - x\frac{\cosh(x)}{\sinh^2(x)}\right\}\frac{\partial x}{\partial T}, \quad (3.43)$$

with

$$x = 2\pi^2 k_B T m^* c / e\hbar H \quad (3.44)$$

and

$$\frac{\partial x}{\partial T} = 2\pi^2 k_B m^* c / e\hbar H. \quad (3.45)$$

Figure 3.10 shows the functional form of both quantities, R_{LK} (blue) and $\partial R_{LK}/\partial T$ (red), as a function of temperature.

While R_{LK} is a monotonically decreasing function with temperature, its derivative $\partial R_{LK}/\partial T$ shows a pronounced maximum.

References

1. Tribus M, McIrvine EC (1971) Energy and information. Sci Am Magaz 225(3):179–184
2. Mandl F (1988) Statistical physics. Wiley Blackwell, New York

3. Gibbs JW (1873) Graphical methods in the thermodynamics of fluids, Transactions of the Connecticut Academy, I. pp. 309–342, April–May
4. Stewart GR (1983) Measurement of low-temperature specific-heat. Rev Sci Instrum 54(1):1–11
5. Sullivan PF, Seidel G (1968) Steady-state, AC-temperature calorimetry. Phys Rev 173(3):679–685
6. Bachmann R, Schwall RE, Thomas HU, Zubeck RB, King CN, Kirsch HC, Disalvo FJ, Geballe TH, Lee KN, Howard RE, Greene RL (1972) Heat-capacity measurements on small samples at low-temperatures. Rev Sci Instrum 43(2):205
7. Hwang JS, Lin K, Tien C (1997) Measurement of heat capacity by fitting the whole temperature response of a heat-pulse calorimeter. Rev Sci Instrum 68(1, Part 1):94–101
8. Lounasmaa OV (1974) Experimental principles and methods below 1 K. Academic Press, London
9. Pobell F (1992) Matter and methods at low temperatures. Springer, Heidelberg
10. Jaime M, Kim KH, Jorge G, McCall S, Mydosh JA (2002) High magnetic field studies of the hidden order transition in URu_2Si_2. Phys Rev Lett 89(28):287201
11. Boyle WS, Hsu FSL, Kunzler JE (1960) Spin splitting of the landau levels in bismuth observed by magnetothermal experiments. Phys Rev 4(6):278–280
12. Condon JH (1966) Nonlinear de Haas-van Alphen effect and magnetic domains in beryllium. Phys Rev 145(2):526–535

Chapter 4
Design and Characterisation of Novel Experimental Setup

The aim of the project presented in this thesis was the study of the entropic properties of $Sr_3Ru_2O_7$ as a function of magnetic field. The sizable magnetic moment of this material as a function of field leads together with the anisotropy of the magnetisation to a significant torque at the metamagnetic phase transitions in the phase diagram of $Sr_3Ru_2O_7$. Furthermore, in order to carry out the magnetocaloric measurements reported here, a high temperature stability and low noise thermometry under magnetic field is necessary. In order to optimise the experiment for these requirements a new setup was designed, built and characterised by the author. In this chapter first a general overview of the design will be given. An important component is the careful calibration of the thermometry as a function of temperature *and* magnetic field, which will be discussed in the second part of this chapter. Finally, characterisation measurements with Sr_2RuO_4 as well as $Sr_3Ru_2O_7$ will be presented.

4.1 Measurement Environment and Sample Holder

All caloric measurements reported in this thesis have been carried out in an Oxford Instruments 'Kelvinox 25' dilution refrigerator fitted with a superconducting 17 T magnet. This instrument is capable of performing measurements down to a base temperature of 25 mK. The operating principle of a dilution refrigerator is widely described in the literature and details can be found for example in [1] or [2]. In the context of this work two points are important to mention. First, the experiment takes place inside an evacuated space immersed in liquid helium. Second, the cooling is achieved in the so-called mixing chamber shown in Fig. 4.1. The name originates from the cooling process which is based on the low temperature properties of the cooling agent—a ^{3}He–^{4}He mixture. As shown in 4.1, this mixing

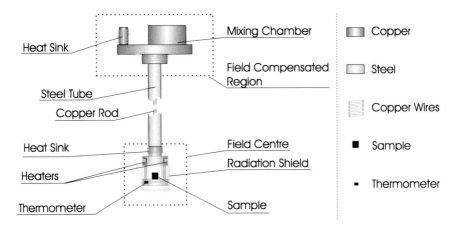

Fig. 4.1 Overview of the experimental environment and the sample holder. Details are given in the text

chamber is installed in a field compensated region.[1] The centre of the magnetic field is approximately 30 cm below it.

In order to position the experimental setup in the field centre an appropriate sample holder had been built. As indicated in Fig. 4.1 it is made of a hollow stainless steel tube with a thin copper rod running along the inside for thermal contact. This combines the structural stability of the steel with the good thermal conductivity of the copper while minimising the possibility for eddy-current heating in the copper during magnetic field sweeps. The tube is terminated by two copper surfaces which connect to the mixing chamber on one end and the experimental setup at the other end. All experimental wiring from the mixing chamber to the experiment is made of twisted pairs of 100 μm insulated copper. The wires are heat sunk at the two positions indicated in Fig. 4.1, which also shows the location of two thin film heaters[2] at the bottom of the probe holder as well as a field and temperature calibrated thermometer.[3] The heaters and thermometer are used to directly stabilise and control the temperature of the experimental setup. Finally, the experiment is shielded by a radiation shield made of 50 μm thin copper foil and screwed into the terminating copper block of the sample holder. This is of particular importance to avoid heating of the thermally isolated sample by thermal radiation from the 4.2 K 'hot' surfaces of the vacuum can.

In the following section, I will describe the experimental setup in the centre of the magnetic field in more detail.

---

[1] One of the main advantages of this design is the possibility of measuring the temperature of the mixing chamber without the complications of thermometry under magnetic field.

[2] These are 500 Ω strain gauges supplied by Kyowa Electronics.

[3] This is a RuO_2 chip thermometer identical in built to the one used as a sample thermometer and described in Sects. 4.2.3 and 4.3.

4.2 Design of Experimental Setup

As mentioned above, the specific heat and magnetocaloric measurements on $Sr_3Ru_2O_7$ under high magnetic fields reported here impose considerable and often competing requirements on the experimental setup. On the one hand, for example, one wants to minimise the relative contribution of the addenda to the total signal and therefore work with the largest samples possible. That, on the other hand, results in the case of $Sr_3Ru_2O_7$ in a large magnetic moment. The resulting forces in magnetic fields can lead in the best case to the sample detaching from the setup and in the worst case to the destruction of the sample stage. In the following section this and other requirements for each of the components will be discussed, explaining the motivation for the particular implementation of the setup chosen in this project. This will be followed by a section giving a rough estimate of the thermodynamic properties of the setup based on the physical properties of the materials used as given in the literature and summarized in Appendix A.

4.2.1 Design Goals and Experimental Realisation

Chapter 3 gave a general introduction to the principles of the thermodynamic measurements used in this work. In particular, (Fig. 3.1) showed a schematic of an idealised experimental situation. This schematic depicted the sample, heater, thermometer and thermal bath as being abstractly linked without any further support. In reality one often uses a sample platform to which one attaches the sample, heater and thermometer. This sample platform is then linked to the thermal bath. The situation appropriate for the experimental design of this work is shown in Fig. 4.2a. Both the sample (S) and the thermometer (Th) are directly attached to the sample platform (SP). The heater (H) is attached to the sample (S). The thermal resistances between these four components are shown in black. From an experimental point of view these resistances have to be minimised in order to achieve thermal equilibrium between the components. In contrast to that, parasitic thermal links in the experiment between the thermometer (Th)/heater (H) and the thermal bath (TB) are shown in red. These are caused by the electrical connections of these components. Their thermal resistance has to be maximised to minimise the amount of heat flowing through them. This parasitic heat flow could cause either the thermometer or heater not to be in thermal equilibrium with the sample, leading to systematic errors. Finally, the thermal link between the sample platform (SP) and the thermal bath (TB) is shown in green. Its thermal conductance k has to be adjustable so that in combination with the total heat capacity of the sample and addenda (C_{total}) the thermal time constant $\tau = C_{total}/k$ is of the correct order of magnitude.

The particular implementation chosen in this project is shown schematically in Fig. 4.2b and as a graphic in Fig. 4.2c. Here, a copper ring acts as the thermal bath.

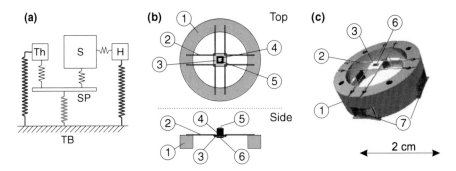

Fig. 4.2 Overview of the experimental setup. **a** Schematic of the thermodynamically relevant components of the setup. These are the sample S, the thermometer Th, the heater H, the sample platform SP and the thermal bath Th. The components are thermally connected as indicated (for details see text). The setup as implemented in this work is shown in **b** and **c**. The main components are a copper ring *1*, Kevlar strings *2*, a silver platform *3*, the sample *4*, the heater *5*, the thermometer *6* and CuBe springs *7*. Not shown are the electrical contacts and the platinum wire for the thermal link

From this a sample platform made of silver is suspended on Kevlar strings. On one side of the sample platform a RuO_2 thermometer is installed whereas the sample is fixed to the other side. The heater is a thin film strain gauge attached to the top of the sample. The electrical contacts are made via high resistance Manganin[4] wires. Finally, the tunable link is achieved via a thin platinum wire that can be adjusted in length and thickness. In the following each of these components will be discussed in detail.

4.2.2 Sample Platform and Thermal Bath

The sample platform and its support structure have been designed to satisfy two main constraints. First of all, from a thermodynamic point of view, the sample platform needs to have a very good thermal conductivity and small specific heat in order to achieve thermal equilibrium between the thermometer and the sample. Second the support structure between the thermal bath and the sample platform has to have an extremely low thermal conductivity to achieve a high tunability of the thermal link.[5] On the other hand the particular experiment of measuring a material with a strong magnetic moment in high magnetic fields requires good mechanical stability.

---

[4] All Manganin wires used in this work are 86%Cu/12%Mn/2%Ni wires of 30 μm diameter supplied by Goodfellow.

[5] One can always increase the thermal conductivity if desired.

Fig. 4.3 a The sample platform (SP) with Kevlar strings and a typical thermometer chip (background is mm paper). **b** The CuBe springs tensioning the Kevlar string

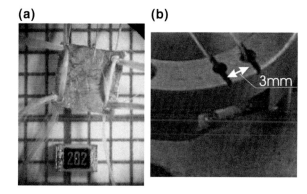

In particular the last condition rules out a standard design where a sample platform is suspended on thin superconducting wires. The design chosen here therefore is to suspend the sample platform on four Kevlar[6] strings as shown in Fig. 4.3a. Each of these strings consists of 35 filaments of a diameter of 17 μm. In order to avoid further addenda contributions by, e.g. epoxy the Kevlar is threaded through holes in the sample platform as shown in Fig. 4.3. The filaments are extremely strong and can be tensioned at room temperature such that the sample platform is fixed in a quasi rigid position.

The platform itself is made of silver.[7] It is approximately 4 mm × 4 mm wide and has a thickness of 0.15 mm. The main consideration here was to use a material of high thermal conductivity that is able to withstand the mechanical stress from the suspension. It furthermore has sufficiently 'soft' edges to avoid damaging the Kevlar strings.

Though the Kevlar strings can be tensioned at room temperature a particular problem is their negative thermal expansion coefficient. This leads to an extension upon lowering of the temperature and without further precautions would cause the sample platform to loosen. Therefore, the Kevlar filaments are fixed with Stycast[8] at the ends to CuBe[9] springs as shown in Fig. 4.3b. This material is elastic to low temperature and keeps the tension of the Kevlar strings. That the platform keeps its rigidity down to 77 K temperatures has been directly tested by immersing it into liquid nitrogen. This is important because most of the total thermal expansion/contraction for materials used takes place between room temperatures and 77 K [3]. A further indirect test is the measured phase diagram of $Sr_3Ru_2O_7$ discussed later (see Sect. 5.1 in Chap. 5).

---

[6] The Kevlar is supplied by Goodfellow.
[7] The material was obtained from Goodfellow and has a nominal purity of 99.5%.
[8] Stycast is an epoxy supplied by DuPont.
[9] The CuBe is a 98% Cu and 2% Be alloy supplied by Goodfellow.

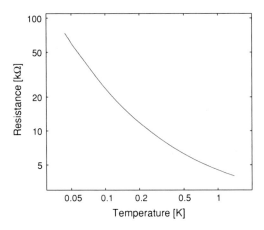

Fig. 4.4 The typical resistance behaviour of a RuO_2 chip as function of temperature

4.2.3 Thermometer

Central to the caloric experiments reported in this work is an exact knowledge of the absolute temperature of the sample over the whole temperature and field region. Since most thermometers have an appreciable dependence on magnetic fields this requires the thermometer to be calibrated as a function of both temperature and field.[10]

The setup presented here uses standard Lakeshore RuO_2 resistance chips[11] with a cross section of 1.45 mm × 1.27 mm and a thickness of 0.3 mm. The resistive material is a thin layer of a RuO_2 based glass deposited on a polycrystalline Al_2O_3 substrate and contacted by platinum pads. In order to reduce the thermal path between the bottom of the substrate and the resistive layer the former was polished down to 100 μm.

Figure 4.4 shows a typical resistance curve as a function of temperature for thermometers of this kind. The exponential increase of resistance towards lower temperature ensures that the sensitivity $d \ln(R)/d \ln(T)$ increases towards lower temperatures. The physical basis for the increasing resistance is the fact that the conductance of the material is dominated by variable range hopping of electrons in the RuO_2-glass matrix. The conductance decreases towards lower temperatures since this hopping is thermally activated.

---

[10] The one type of sensor I am aware of that is field independent is based on the dielectric constant of certain glasses. However these have the problem of long times to reach thermal equilibrium and are in their current form not suitable for the experiment.

[11] Model RX-202

4.2 Design of Experimental Setup

In the experiment the thermometer is attached to the silver platform via Apiezon grease. The electrical contacts for the four-point resistance measurement were made with twisted pairs of Manganin wires with a diameter of 30 μm. These wires could not be attached by solder since heating the thermometers significantly above room temperature would have destroyed their calibration. It was therefore decided to make the contact with a silver based conductive epoxy.[12] Since these contacts are not very strong mechanically they furthermore had to be stabilised with a minimal amount of Stycast.

The details of the thermometer calibration and characterisation carried out for this thesis are discussed in Sect. 4.3.

4.2.4 Heater

The final component of the experimental setup is the heater. As in the case of the thermometer it has to satisfy the condition of having a small thermal mass. Furthermore, one would like to have a resistive element that does not significantly change its resistance as function of temperature or magnetic field. Here, it was decided to use thin film strain gauges. During the course of the experiment two different strain gauges were used, The first is a 120 Ω heater supplied by Kyowa Electronics.[13] The substrate is a very thin polyimide resin of approximately 13 μm thickness. The resistive element of the gauge is a CuNi alloy film. The mass is less than 1 mg.

The second type a 5 kΩ strain gauge supplied by Vishay.[14] Here the film is made of a modified K-alloy (NiCr) on a glass–fibre-reinforced epoxy substrate[15] of approximately 13 μm thickness and a total weight of less than 1 mg.

For the experiment one of these heaters is attached to the sample via conductive silver epoxy[16] in order to have optimal thermal contact. Furthermore, they are contacted by two twisted pairs of Manganin wires of 30 μm diameter to supply the heater current and being able to simultaneously carry out a four point measurement of their resistance. Over the full temperature (100 mK to 1.5 K) and magnetic field range (0–15 T) the heater resistance was found to vary less than ±0.25%.

---

[12] DuPont 4929 room temperature silver paste.
[13] Model KFG02-120-C1-16
[14] Model SK-06-S022H-50C
[15] As a technical aside it has been found that the solvent used in GE varnish destroys this substrate.
[16] DuPont 4929

4.2.5 Estimates of Thermal Performance

In order to have a guide for the experiment the main thermal characteristics of the setup were estimated. The bases for this are the physical properties of the materials used as found in the literature. A detailed list of the experimental values as well as the original sources is given in Appendix A. Here, the main results for the estimate of the heat capacity of the addenda, the thermal link to the bath, as well as the resulting thermal time constants are given. It should be emphasised that the intention is not to calculate exactly the properties of the setup but rather to provide a guide for the rough order of magnitude of the properties of the setup as well as possible variations with temperature and magnetic field.

4.2.5.1 Addenda Heat Capacity

The specific heat of almost all of the materials used for the experimental setup has been measured previously. Figure 4.5 shows the most significant contributions to the heat capacity C_{add} of the addenda. It is given as C_{add}/T as a function of temperature to be more clearly comparable with measurement results presented later, which are also shown divided by temperature. The shaded regions indicate the contributions of each component leading to the total heat capacity shown by the red curve. These contributions are based on an estimate of the total amount of material in the setup in combination with the literature specific heat values.

The first main contribution (1) comes from the silver platform. Here, the mass of the platform was measured to be 17.3 mg and the source for the specific heat is du Chatenier and Miedema [4]. The physical origin for the heat capacity of the platform at these temperatures is primarily the electronic specific heat of this metal.

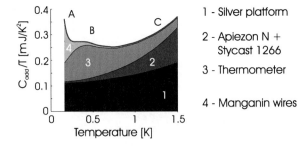

Fig. 4.5 Here, the red curve shows the addenda contribution to the heat capacity of the setup as a function of temperature. The shaded areas under the curve correspond to the contribution of different components (*1*, *2*, *3* and *4*). Furthermore three significant features in the background contribution are emphasised by *A*, *B* and *C*. Further details are given in the text

4.2 Design of Experimental Setup

The second contribution (2) is coming from amorphous materials such as the Apiezon grease (measured by Schink and Löhneysen [5]) used to fix the thermometer to the platform and Stycast (measured by Siqueira and Rapp [6]) that is used to fix the electrical contacts to the heater and thermometer. Since the amount of amorphous material applied is not directly measurable it was assumed here that both components have a mass of less than 1 mg. Furthermore, the specific heat at low temperatures is of the same order of magnitude and has a similar temperature dependence for almost all amorphous materials [7]. The contribution indicated by region (2) can therefore also be understood as a sum total of all amorphous components[17] of a total mass of the order of 1 mg.

Component (3) is an estimate of the heat capacity contribution by the thermometer itself. Unfortunately this particular thermometer has not been measured previously. However Volokitin et al. [8] have investigated the heat capacity of a range of similar resistance chips. The common feature of all was a peak in the heat capacity in the range from 0.3 to 0.7 K. Here, the behaviour of one such resistance chip that is most comparable to the dimensions and resistance value of the one used in this work is shown.

Finally, contribution (4) gives the heat capacity of the Manganin wires that are directly fixed and thereby thermally anchored to the thermometer with Stycast.[18] While the phonon and electronic contribution is negligible, Manganin shows a significant low temperature Schottky anomaly. The data presented here is based on the work by Ho et al. [9]. As in the case of amorphous materials, a Schottky anomaly is a generic feature of magnetic impurities or nuclear magnetic contributions. The contribution shown here is therefore to be understood as being representative for the amplitude of a Schottky anomaly at low temperatures that can be caused by any number of magnetic impurities in the materials used in this experiment.

Finally it should be mentioned that for the first three contributions the magnetic field dependence is not known in detail but can be expected to be weak. Contrary to this it is known that the 'high' temperature tail of a magnetic Schottky anomaly as shown here scales with temperature T and magnetic field H as H^2/T^2.

Overall these four contributions lead to a total heat capacity as indicated by the red curve. In particular, attention should be drawn to the three features in the temperature dependence indicated by A, B and C. These are predominantly caused by a Schottky anomaly in the Manganin wires (A), the peak in the heat capacity of the thermometer (B) and by the contribution of amorphous components (C). As will be seen later two of these features can be clearly identified in the experimentally determined addenda contribution.

---

[17] This includes small possible additions from the Kevlar strings and the support of the heater.

[18] This is a total length of 12 mm wire of 30 μm diameter. Wire that is not directly glued to the thermometer is assumed to be thermally inert on the timescale of the experiment due to the high thermal resistance of this alloy.

4.2.5.2 Thermal Conductances

The second characteristic property of the setup is the thermal link between the sample platform and the copper ring acting as the thermal bath. There are three materials connecting these two.

First of all there are eight strings of Kevlar each of which consists of 35 filaments of 17 μm diameter and 8 mm length. The thermal conductivity κ_{Kevlar} of Kevlar has been measured by Ventura et al. [10] and found to be well described by

$$\kappa_{Kevlar} = 3.9 \times 10^{-5} T^{1.17} \text{ W/cm K}$$

over a range of temperature T from 0.1 to 2.5 K.

The second contribution to the thermal conductivity comes from the electrical contacts to the thermometer and heater. These are four twisted pairs of Manganin wire with a diameter of 30 μm per wire and a length of approximately 1.5 cm. In the work by Peroni et al. [11] the thermal conductivity $\kappa_{Manganin}$ has been found to follow a similar power law of

$$\kappa_{Manganin} = 9.5 \times 10^{-4} \times T^{1.19} \text{ W/cm K}.$$

The final thermal link is that made by a 30 μm thick and 1 cm long platinum wire. Its thermal conductivity was reported by Davey et al. [12] to be

$$\kappa_{Platinum} = 0.32 \times T \text{ W/cm K}.$$

The thermal conductance k is given by

$$k = \kappa \frac{A}{L}, \tag{4.1}$$

with A being the cross section and L the length of the thermal link. This gives for our three components at $T = 1$ K the following three values:

$$\begin{aligned} k_{kevlar,1K} &= 3.1 \times 10^{-8} \text{ W/K}, \\ k_{Manganin,1K} &= 3.6 \times 10^{-8} \text{ W/K}, \\ k_{Pt,1K} &= 2.2 \times 10^{-6} \text{ W/K}. \end{aligned} \tag{4.2}$$

A word of caution here should be that the value for platinum had been determined for high purity single crystalline material whereas the wire used as a thermal link in the experiment was not annealed. This can cause the thermal conductivity to be significantly reduced. Experimentally one finds a value which would be expected to be of the order of $k_{1K} = 5 \times 10^{-7}$ W/K with the temperature dependence being close to the expected $k(T) \propto T$.

4.2.5.3 Thermal Time Constants

Finally, I would like to discuss the typical thermal time constants to be expected from the experimental setup. The sample platform to thermal bath relaxation time constant τ is given by

4.2 Design of Experimental Setup

$$\tau(T) = \frac{C_{\text{add}}(T) + C_{\text{sample}}(T)}{k(T)} \quad (4.3)$$

One would therefore expect without any sample on the setup an addenda-only relaxation time of the order of 0.6 s assuming an average heat capacity of $3 \times 10^{-7} T$ J/K and a thermal conductance of $5 \times 10^{-7} T$ W/K. A typical sample of $Sr_3Ru_2O_7$ of a total weight of 20 mg on the other hand has in zero field a heat capacity of the order of $7.5 \times 10^{-6} T$ J/K giving a typical time constant for the experiment of 15 s.

So far the design as well as theoretical performance of the setup have been discussed. In the final two sections of this chapter experimental results describing its characteristics will be presented. All of these measurements however depend crucially on the thermometer installed on the sample platform being calibrated as function of temperature as well as magnetic field. Therefore, the next section first discusses the details of the thermometer calibration.

4.3 Thermometry

As discussed earlier a central part of the specific heat setup is the thermometry. For this work RuO_2 based thick film resistance chips have been chosen as thermometers and their general properties described in the previous section. Here, the known electrical properties of RuO_2 chips in the low temperature range will be reviewed first. In the second part the details of the calibration are presented.

4.3.1 Physical Properties of Thermometers

4.3.1.1 Low Temperature Resistance

RuO_2 resistive chips have been used for thermometry for over 20 years. Their main advantage is the steep rise in electrical resistance R upon lowering the temperature T. However each individual resistor has a slightly different resistance as a function of temperature requiring individual calibration. In Fig. 4.6 the resistance curve is shown for the calibrated thermometer that was used in this work as the temperature standard.[19]

An important quantity for thermometers is their sensitivity, meaning the relative change in resistance upon a relative change in temperature. This quantity is given by $d(\ln R)/d(\ln T)$. The log–log plot in Fig. 4.6 shows that this quantity is relatively constant over the temperature range of interest to this experiment.

---

[19] The calibration was done by Lakeshore and the serial number of the thermometer is U01518.

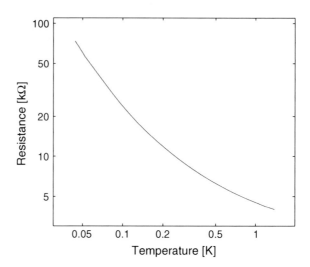

Fig. 4.6 Low temperature resistance R as a function of temperature T for the calibrated Lakeshore RuO_2 chip U01580 used as the temperature standard for this work

The above calibration by Lakeshore had been performed on 77 logarithmically spaced temperature points between 43 mK and 45 K and has been interpolated in the three temperature regions from base temperature to 800 mK, from 800 mK to 6 K, and from 6 to 40 K by Chebychev polynomials of degree ten. These have the form of

$$T = \sum_{i=0}^{i=10} A_i \cos(i \arccos(X)), \quad (4.4)$$

where T is the temperature, A_i are coefficients provided by Lakeshore and X is given by

$$X = \frac{(\log_{10}(R) - \log_{10}(R_{\text{low}})) - (\log_{10}(R_{\text{high}}) - \log_{10}(R))}{\log_{10}(R_{\text{high}}) - \log_{10}(R_{\text{low}})}. \quad (4.5)$$

Here, R is the resistance and R_{low} and R_{high} the resistance values at the lower and upper end of the interpolation range. This interpolation scheme was used during the calibration measurements to determine the temperature of the calibrated thermometer from its measured resistance.

4.3.1.2 Magnetoresistance

Of particular interest in this thesis is the performance of the developed setup under magnetic fields. The magnetoresistance of RuO_2 chips has been measured by several authors. In Fig. 4.7 the results by Goodrich et al. [13] are presented. Here the relative change of the resistance R as a function of field is shown in a range from 0 to 18 T. The temperature is 28 mK.

4.3 Thermometry

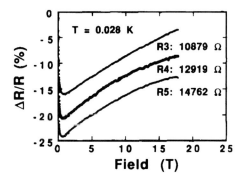

Fig. 4.7 Here the relative change in resistance $\Delta R/R$ as a function of magnetic field is shown for several individual resistors [14]. The temperature during the experiment was 28 mK

The resistance initially drops and goes through a minimum at around 1 T. Above this field the resistance starts raising monotonically. It was shown in the same paper that the change in resistance above 3 T is well described by a linear dependence on $\mu_0 H^{1/2}$ with H being the applied field. However the particular proportionality coefficient is temperature dependent, requiring its measurement over the full temperature range.

Finally, it has been shown that the absolute resistance of RuO_2 chips can change by thermal cycling, which would in principle make a thermometer calibration impossible. It was shown however in [14] that the resistance saturates as a function of thermal cycles. The results of that study are shown in Fig. 4.8. Here in each cycle the five resistors indicated are cooled from room temperature to 4.2 K and their resistance measured. A resistance difference of 10 Ω would correspond to a temperature error of approximately 65 mK or 1.5%. The changes are saturating after approximately 60 cycles. The physical origin of the effect has been ascribed to structural changes (i.e. micro cracks) due to mechanical stresses in the resistance chips.

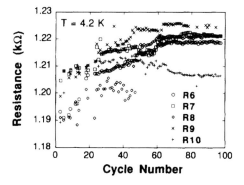

Fig. 4.8 Resistance as a function of thermal cycles to 4.2 K for several resistors [13]

4.3.2 Thermometer Calibration

Having described the relevant resistive properties of RuO_2 chips in the previous section, the calibration procedure itself is detailed here. A total of four chips were calibrated. They are all model RX-102 bare chips obtained from Lakeshore.

4.3.2.1 Thermal Cycling

As mentioned above mechanical stress can alter the resistive properties of the resistive chips. Therefore, they had to be thermally cycled after removing part of the substrate by polishing (see Sect. 4.2.3). This was done with the resistivity probe of a Quantum Design MPMS system. This system allows the programming of temperature cycles between room temperature and 4 K while providing the possibility of monitoring the resistance during the process.

Figure 4.9a shows the resistance of two of the RuO_2 chips as a function of cycle number. As one can see there are variations of the absolute resistance that are correlated with each other. It is assumed that these variations come from systematic errors in the temperature stability of the MPMS system. Already, however, it is possible to determine that the absolute changes are below 0.5%. A better way for looking at resistance fluctuations with thermal cycles is to look at the relative ratio between two of the resistors and thus eliminating to first order the systematic temperature error. This is shown in Fig. 4.9b. Here it is apparent that after 40 cycles the resistance changes are of the order of 0.1%. It was therefore assumed that the resistors were sufficiently cycled.

4.3.2.2 Low Temperature Calibration Probe

For the calibration itself in the previously mentioned Oxford Instruments cryostat a new calibration probe was developed. It is shown schematically in Fig. 4.10.

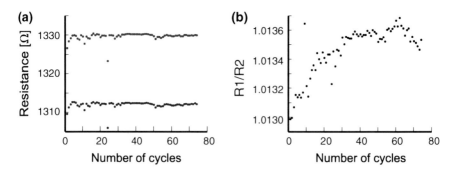

Fig. 4.9 **a** Resistance of two RuO_2 thermometer chips as a function of thermal cycles with base temperature 4 K. **b** The relative resistance of these two chips as a function of thermal cycle in order to eliminate systematic temperature fluctuations of the Quantum Design PPMS system

4.3 Thermometry

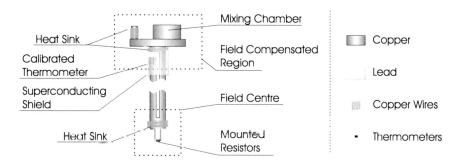

Fig. 4.10 Schematic overview of the calibration probe. Details are given in the text

The thermometers to be calibrated were installed at the bottom of the probe in the centre of the magnetic field. The probe design is primarily driven by the fact that the calibrated thermometer is a RuO_2 chip in itself. To avoid any magnetoresistive effects it had to be installed in the field compensated region at the mixing chamber. Here the magnetic field is smaller than 200 Gauss. To reduce the effect further, the top of the probe was surrounded by a superconducting shield made of lead formed into the shape of a tube.

In order to make sure that the calibrated thermometer is measuring the temperature of the resistors rather than the temperature of the mixing chamber the probe was slit along its length and all wires leading to the calibrated thermometer were carefully heatsunk at the same positions as those of the uncalibrated resistors.

Figure 4.11a shows a schematic of the electrical wiring. On the left hand side the current source is shown. For all experiments reported here a Keithley 6221 AC current source was used. The blue region in the centre indicates the part of the electronics that is inside the cryostat, which is only the measured resistance R and the two twisted pairs for current and voltage leads of the 4-point resistance measurement. On the very right the lock-in amplifier used as a volt meter in this experiment is shown with its effective input impedance. All measurements reported in this work were performed with Stanford Research SR830 lock-ins. The frequency signal for the lock-in process was supplied via an external TTL trigger line from the Keithley current source.

One problem in low temperature thermometry is high frequency noise. While not affecting the lock-in amplifier measurement this noise can cause the resistance thermometers to be internally heated by Joule heating. In order to avoid this, high frequency filters were installed on all lines with the component values given in Fig. 4.11a. The performance for this filter was simulated for a typical resistance value of the thermometer chip of 20 kΩ. In Fig. 4.11b the amplitude and phase shift of the voltage signal as seen by the lock-in amplifier is shown as function of frequency of the AC current. The -3 dB value usually defined as the cut off frequency for a filter is reached at around 20 kHz. This results in a signal loss of less than 0.1% below 20 Hz. After a careful study of amplitude versus frequency for a given excitation current it was decided to use 13.1 Hz as the excitation

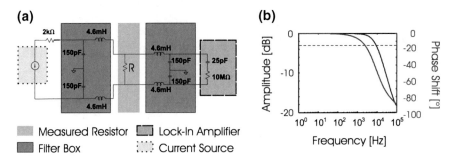

Fig. 4.11 a The electric circuit used for all thermometry measurements. The current source and lock-in amplifier are both at room temperature outside the cryostat. Directly at the cryostat filter elements of the given characteristics are installed in a filter box. The blue region marks the part of the electronics inside the cryostat, i.e. the RuO_2 chip thermometer and wiring for a four point measurement. **b** The voltage signal (*blue*) in dB as measured across the effective input impedance of the lock-in amplifier together with the phase shift (*red*) as a function of frequency of the excitation current. 0 dB corresponds to full signal whereas the -3 dB value marked by the dotted line defines the cut-off frequency of the filter

frequency for all thermometry measurements. The excitation current was chosen to be 0.3 nA below 100 mK and 2.5 nA above 100 mK to stay within the recommended regions of power dissipation in the RuO_2 chips in order to avoid self heating of these resistors by the measurement current. It was experimentally verified that this value is well below the excitation current above which self-heating effects become measurable.

All four resistors as well as the calibrated thermometer were driven by the same current source with the voltage measurement being done with five lock-in amplifiers operating in parallel.

4.3.2.3 Calibration Procedure

Figure 4.12a shows a single step of the calibration measurements. The top part shows the temperature of the calibrated Lakeshore thermometer as a function of time t whereas the lower part shows the resistance of one of the RuO_2 chips on the same timescale. At $t = 0$ the temperature set point is changed from 98 to 102 mK. Both the thermometer and the chip stabilise on the same timescale which rules out any significant thermal resistance between them. Over the range where both thermometer and resistor have reached stability the temperature and resistance values are averaged. This average value constitutes a single point in the resistance versus temperature graph shown in Fig. 4.12b for one of the RuO_2 chips. The measurement points are spaced in 5% steps between 50 mK and 1.5 K resulting in a total of approximately 70 data points (twice as many as the Lakeshore calibration over the same temperature range). The gap in data points around 1 K is caused by a thermodynamic instability of the $^3He - ^4He$ mixture of the cryostat making a sufficiently good temperature stabilisation impossible.

4.3 Thermometry 81

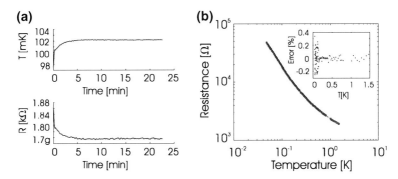

Fig. 4.12 **a** A single step of the calibration procedure. In the *upper graph* the temperature of the calibrated RuO$_2$ chip is shown as a function of time. The *lower graph* gives the resistance of one of the uncalibrated thermometer chips over the same time. The resistance and temperature for a single calibration point is measured as the mean value over the region where the signals are stable. **b** The resulting calibration curve of the resistance R as a function of temperature T (*blue points*). The *red* curve is an interpolating smoothing spline through this data. In the *inset* the residuals of this fit are shown

The red curve in Fig. 4.12b shows the interpolating smoothing spline that was used to parameterise the data over the given temperature range. The inset shows the relative deviation of the measured resistance of the data points R_{measured} from the interpolating spline $R_{\text{interpolated}}$ as given by Error $= (R_{\text{measured}} - R_{\text{interpolated}})/R_{\text{interpolated}}$ as a function of temperature.

In the region above 100 mK where an excitation current of 2.5 nA has been used the error distribution is smaller than 0.1%. Below 100 mK this error distribution increases significantly mainly due to the lower excitation current of 0.3 nA.

The above described calibration was done at 0 T as well as at 11 fields between 3.5 and 15 T that were evenly spaced as a function of the square root of the magnetic field H, i.e. evenly in $H^{1/2}$ in order to reflect the expected behaviour of the magnetoresistance of the RuO$_2$ chips. From the data gathered, a surface of the form $a(T) + b(T)H^{-1/2}$ was fitted to the quantity $(R(\mu_0 H) - R(0T)/R(0T))$ for each of the resistance chips with $a(T)$ and $b(T)$ being fit parameters at each temperature T. The resulting fit surface as a function of magnetic field and temperature is shown in Fig. 4.13a together with the data points used for the calibration. Figure 4.13b shows the relative deviation of all measurement points from the fit surface. The error is less than 0.5% for all points except a few outliers.

The importance of calibrating the thermometer both as a function of temperature and magnetic field is stressed in Fig. 4.14. Here the temperature error that one would make if one would use the zero field calibration for all fields is shown. The blue line indicates the thermometer temperature at 100 mK whereas the red curve is the temperature one would believe the thermometer to be at if using the zero field calibration—an error of the order of 8–16%.

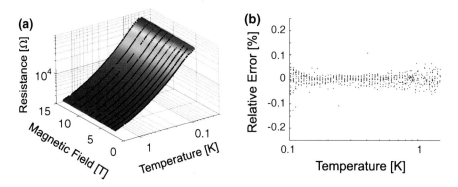

Fig. 4.13 a In *black* are shown the calibration points for the resistance of one of the thermometer chips both as a function of temperature and magnetic field. The surface is calculated via the described interpolation scheme. **b** The relative deviation ($R_{meas} - R_{calc}/R_{calc}$) of the measured resistance R_{meas} from the calculated resistance R_{calc} as a function of temperature for all magnetic fields

4.4 Characterisation Run With Sr_2RuO_4

In order to test the performance of the setup we chose to measure the well established compound Sr_2RuO_4. There are several reasons for this. First of all the compound has a specific heat and thermal conductivity sufficiently similar to $Sr_3Ru_2O_7$ that a realistic test under the final experimental conditions is possible. Second, ultra high purity samples of Sr_2RuO_4 have, as discussed in Chap. 2, several interesting thermodynamic properties in the temperature-field region the setup has been designed for. First high quality Sr_2RuO_4 has in zero field a phase transition to superconductivity at 1.5 K. This allows us to put a lower bound on the intrinsic measurable width of a phase transition. Secondly it has been shown that its specific heat above the critical field for superconductivity is Fermi liquid like and independent of magnetic field. The availability of sufficiently large high

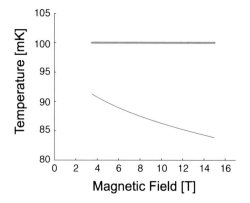

Fig. 4.14 The figure shows the temperature one would wrongly assume (*red curve*) as a function of magnetic field with the thermometer being at 100 mK (*blue curve*) if one does not take into account the field dependent corrections to the calibration

quality single crystal samples in our group particularly allowed to study the possibility of measuring magnetothermal oscillations that occur in the Fermi liquid state.

In this section I will first discuss the relevant properties of the Sr_2RuO_4 sample used in this project. This is then followed by a presentation of the characterisation measurements themselves.

4.4.1 Sample

The sample used in this experiment was cut from a single crystal grown and characterised by A. S. Gibbs in the group of Prof. A. P. Mackenzie. The residual resistivity of the crystal was measured to be 0.12 μΩ cm, corresponding to a mean free path of 3,000 Å. Figure 4.15 shows the superconducting transition in zero field as measured by the out of phase component χ'' of the AC magnetic susceptibility as a function of temperature. Mackenzie et al. [14] have shown that the transition temperature is strongly dependent on impurities present in Sr_2RuO_4 samples. The transition temperature of 1.49 K as given by the peak position in the AC susceptibility places this crystal among the best grown anywhere in the world.

For the experiments reported here an approximately 1.5 mm × 1.5 mm × 3 mm large parallelepiped with a total sample mass of 37.8 mg was used.[20] One of the two square shaped surfaces of the sample was cleaved along the crystallographic *ab*-plane. This allowed for accurate mounting of the sample with the *ab*-plane parallel to the surface of the silver sample platform and therefore perpendicular to the magnetic field.

4.4.2 Specific Heat in Field

First the heat capacity under magnetic field was investigated in order to establish the addenda contribution. Here two of the measurement methods described in Chap. 3 were used—the relaxation time as well as the AC heating method.

Figure 4.16a shows an example measurement for the relaxation time method at 410 mK and a magnetic field of 4 T. Here the temperature change of the sample relative to the temperature at time $t_0 = 0$ is plotted as a function of time. At t_0 an AC current of 0.85 μA and a frequency of 100 Hz is applied to the heater. With a resistance of 4.995 kΩ this results in a power dissipation of 3.6 nW. This causes the sample to heat up by 12 mK. The green line marks the range over which the average temperature before the heating step was measured. The red curve is an exponential fit to the data at times $t > t_0$. From this one extracts a time constant τ,

---

[20] This corresponds roughly to 1.11×10^{-4} Ru−mol.

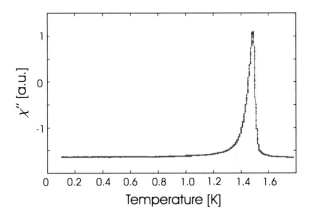

Fig. 4.15 Out of phase component of the AC magnetic susceptibility χ'' as a function of temperature. Graphic courtesy of A. S. Gibbs

typical for all measurements presented, of $\tau = 6.3$ s. As described earlier the temperature step ΔT in combination with the applied heat power $\dot{Q}$ gives the thermal conductance k between the sample platform and the thermal bath as $k = \dot{Q}/\Delta T$. One therefore can calculate the total heat capacity of the setup C as $C = k\tau$ with τ. This has been done for a range of 11 temperatures between 200 mK and 1.2 K. At each temperature the procedure was carried out ten times and the results averaged. The heat capacity C divided by temperature T obtained this way is shown in blue in Fig. 4.16c.

The second method used is the AC heating method. Figure 4.16b shows in blue the temperature trace of the sample while a continuous 40 mHz AC current is applied. The AC measurement starts at a thermal bath temperature of 1.2 K with a current of 4.25 µA. During the experiment both the temperature of the thermal

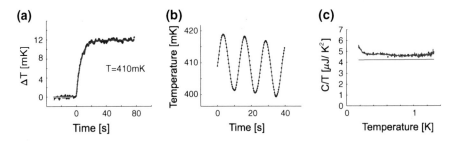

Fig. 4.16 Part **a** Shows the temperature difference ΔT of the sample and addenda as a function of time for a typical relaxation time measurement. From time $t = 0$ continuous heating is applied to the setup. The *green line* shows the fit through the data in order to establish the initial temperature whereas the *red curve* is an exponential fit to the data to times above $t = 0$. Part **b** Shows an analysis step for the AC measurement method. Here in blue the temperature of the setups is shown as a function of time while AC heating is applied and the temperature of the thermal bath is slowly lowered. The *red curve* shows a fit to the data as explained in the text. Part **c** Finally shows the result for the heat capacity divided by temperature as a function of temperature for both the relaxation time method (*blue*) and the AC method (*red*) (further details are given in the text). The *black curve* gives the expected heat capacity of the sample according to literature values [15]

4.4 Characterisation Run With Sr$_2$RuO$_4$

bath and the current are lowered linearly at an approximate rate of 3.9 mK/min and 2.7 nA/min respectively until base temperature is reached. This causes sinusoidal temperature oscillations as a function of time on the background of a linearly decreasing temperature, as shown in the data presented. The frequency f of these oscillations is twice the excitation frequency or $f = 80$ mHz. The amplitude A of the oscillations is extracted from the data by fitting the temperature signal over time intervals of 45 s with a function of the form

$$T(t) = T_0 + at + A\sin(2\pi f + \phi_0), \qquad (4.6)$$

where T_0, a and ϕ_0 are free parameters of the fit. Such a fit is shown in Fig. 4.16b in red. As was discussed in the previous chapter the heat capacity C of the sample is then proportional to the inverse of the amplitude A^{-1}.

In Fig. 4.16c the heat capacity C of the setup divided by temperature T is shown as a function of temperature as obtained by the relaxation time (blue) and AC heating method (red).[21] The black curve represents the heat capacity divided by temperature of the Sr$_2$RuO$_4$ sample as calculated from literature values. The difference between these is the heat capacity of the addenda. It is plotted in Fig. 4.17a together with the background contribution to the heat capacity at 14 T. Here one can identify three characteristic features labelled (A), (B) and (C). As discussed in Sect. 4.2.5 these can be attributed to a Schottky anomaly (A), the thermometer(B) and a contribution from amorphous materials (C).

The high temperature addenda contribution C_{add}/T for both 4 and 14 T is quasi identical whereas at low temperatures they deviate. Figure 4.17b shows the difference $\Delta C(T) = C_{add,4\ T}/T - C_{add,14\ T}/T$ between these two curves together with a fit of the form $\Delta C(T) = a/(T^2)$ which corresponds to the expected functional form of the change in a magnetic Schottky contribution to the heat capacity between different magnetic fields. Since the fit is a very good representation of the data, it is concluded that the heat capacity of the addenda is independent of magnetic field besides a Schottky contribution. This is a significant finding since it shows that the entropy of the addenda above the characteristic onset of the Schottky anomaly at 250 mK can be considered constant. Therefore, any measured entropy changes in the magnetocaloric experiments presented in Chap. 5 have to come from the sample.

Compared to the absolute value estimated in Sect. 4.2.5 the field independent contribution is a factor 1.5 to 2 larger. Taking into account the uncertainties in the theoretical estimate this can be considered as a good agreement, in particular since they agree qualitatively.

---

[21] As described in 3 the AC heating method though having a better relative accuracy than the relaxation time method tends to systematically overestimate the heat capacity for several reason such as finite sample dimensions. Here a multiplicative correction factor of 0.95 has been applied in order to minimise the mean square deviation of the AC heating data from the more absolute accurate relaxation time measurement.

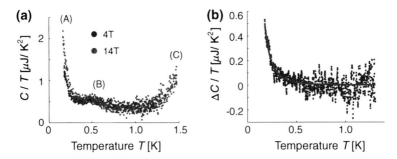

Fig. 4.17 a The addenda contribution to the measured heat capacity as a function of temperature for two different magnetic fields as indicated. The expected features of a Schottky anomaly A, the thermometer B and the contribution from amorphous materials C are clearly identifiable. **b** The difference in addenda contribution between 14 and 4 T. This shows that only the Schottky anomaly is field dependent. The *red curve* is a fit through the data

4.4.3 Specific Heat at Zero Field

After having established the addenda contribution to the heat capacity, the specific heat c in zero field was measured as a function of temperature. The results, corrected for the addenda contribution, are shown in Fig. 4.18 (blue curve) together with the addenda corrected results at 4 and 14 T. In red, data from the literature obtained by Nishizaki et al. [16] are shown.

The data taken here is in very good qualitative and quantitative agreement. The jump at the phase transition is resolved to within 45(5) mK.

4.4.4 Magnetocaloric Oscillations

Finally, due to the high quality of the sample, it was possible to study magneto-thermal oscillations. As discussed in Chap. 3 these temperature oscillations which

Fig. 4.18 Here the measured heat capacity of the sample in zero field (*blue*) is compared with the literature data (*red*) [16]. In *green* and *black* the specific heat at 4 and 14 T is given for comparison

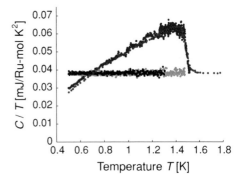

4.4 Characterisation Run With Sr$_2$RuO$_4$

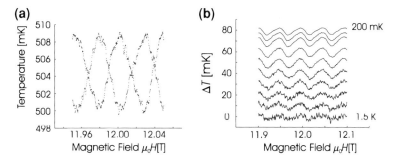

Fig. 4.19 a The observed temperature oscillations for an upsweep and downsweep of the magnetic field in *blue* and *red*, respectively. **b** The temperature oscillations of the sample relative to the bath temperature are shown for a number of thermal bath temperatures (the signals have been offset for clarity)

occur while sweeping the magnetic field are caused by oscillation in the entropy of the sample as a function of magnetic field and are a further manifestation of the same effect causing de Haas–van Alphen oscillations.

Figure 4.19a shows the temperature of the sample during an upward (downward) sweep of the magnetic field in blue (red).[22] The sweep rate chosen was + (−)0.05 T/min. The oscillations are clearly identifiable and their relative phase is 180° as one expects from a magnetocaloric signal if the sample is coupled strongly to the thermal bath.[23]

An important characteristic of quantum oscillations is their temperature independent phase. Figure 4.19b shows the temperature variation of the sample as a function of magnetic field during field sweeps with the same sweep rate at the temperatures indicated. The data for different temperatures has been offset for clarity. Here one can clearly see that the maxima and minima of the oscillations do not change position. Furthermore, it is possible to identify a clear temperature dependence of the amplitude $A(T)$ of the oscillations. As previously shown (see Chap. 3) it should be proportional to the temperature derivative of the well known Lifshitz–Kosevich formula.

In order to analyse the temperature dependence in more detail the Fourier transformation of the oscillatory signal between 11.5 and 12.5 T as a function of inverse field was taken for all temperatures shown.

In Fig. 4.20a the frequency spectrum for one such temperature ($T = 500$ mK) is given together with a Gaussian fit to its main peak at 2,975 T, This frequency agrees well with the previously published result of 3.0 kT [17].

---

[22] The curves are corrected for a known hysteresis in the magnetic field sweeps of ± 15 mT.

[23] In particular this phase shift rules out that the oscillations are oscillations in parasitic eddy-current heating due to Shubnikov–de Haas oscillations in the resistivity. These would show a phase shift of 0° due to the fact that eddy-current heating does not depend on the sign of the sweep rate.

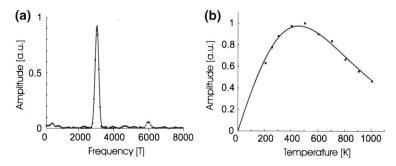

Fig. 4.20 **a** Shows a frequency spectrum of the observed magnetothermal oscillations (*blue*) together with a Gaussian fit to the main peak (*red*). **b** Shows the temperature dependence of the amplitude of the oscillation together with a fit of the temperature derivative of the Lifshitz–Kosevich formula

Figure 4.20b shows the amplitude of the oscillations as a function of temperature as determined from the area under the Gaussian fit to the frequency spectrum. The red curve is a fit of the temperature derivative of the Lifshitz–Kosevich formula giving a mass of $(3.0 \pm 0.1)\, m_e$. This is slightly lower that the literature value of $3.3\, m_e$ but in good agreement with other experiments conducted on recent samples [18].

4.5 Characteristics and Details of Measurements on $Sr_3Ru_2O_7$

4.5.1 Sample Selection

The samples used in this experiment were grown and characterised in a previous study [19]. Here two samples of that study have been used. Their quality has been determined by several characterisation measurements, the results of which are summarised in Table 4.1. One of the main criteria was a low residual resistivity of the order of $0.5\,\mu\Omega$ cm. Furthermore the content of impurity phases of

Table 4.1 Summary of characterisation measurements for samples C698J and C698K

Property	C698J	C698K
Mass	24.1 mg	23.0 mg
Residual resistivity ρ_0	0.48 $\mu\Omega$ cm	0.51 $\mu\Omega$ cm
Normalised dHvA amplitude	11.6 a.u.	31 a.u.
Impurity Sr_2RuO_4	0.6%	2%
Impurity $SrRuO_3$	0.008%	0.03%
Impurity $Sr_4Ru_3O_{10}$	0.04%	0.05%

The mass is the sample mass used in the experiments here. All other data is reproduced from [19]. For details see text

Sr_2RuO_4, $SrRuO_3$ and $Sr_4Ru_3O_{10}$ as determined from AC magnetic susceptibility studies had to be small.

The main sample on which the magnetocaloric and specific heat experiments were performed is C698J due to its higher purity. It was cut to rectangular dimensions with an *ab*-plane cross section of approximately 1.5 mm × 1.5 mm and a height of ≈1 mm in the crystallographic *c* direction. The second sample C698K was a control sample for reproducibility of the experiments. Because of its higher de Haas-van Alphen amplitude it was also better suited to the measurement of magnetocaloric oscillations. Therefore, though the effect has been observed in both samples, oscillation measurements reported here are only those for C698K. Furthermore, this sample has been optimised for an initial angular dependence study of the magnetocaloric signal by being cut in a specific way described in detail in Appendix B.

4.5.2 Thermal Link

An important characteristic of the experimental setup is the thermal link k between the sample platform and the copper ring acting as the thermal bath. In order to know the thermal conductance k both as a function of magnetic field H and temperature T it has been measured in situ during the main experiments on sample C698J. The measurement was performed by applying a constant heat input $\dot{Q}$. Together with the measurement of the sample temperature with heat input T_S and the temperature of the sample without applying additional heat T_0 it is then possible to calculate k via

$$k = \frac{\dot{Q}}{T_S - T_0}. \tag{4.7}$$

In Fig. 4.21a three traces of the thermal conductance k divided by temperature T are shown as a function of magnetic field H. The temperature of the setup during each of these sweeps is $T_{setup} = 200$ mK, $T_{setup} = 500$ mK and $T_{setup} = 1200$ mK as indicated. One can observe a decrease in thermal conductance as a function of magnetic field that can be described by a linear fit within the resolution of the experiment as indicated in red.

Figure 4.21b on the other hand shows a trace of the thermal conductance k divided by temperature T as a function of T. The applied magnetic field is 8 T. The thermal conductance is not a simple constant as expected from a pure metallic link. The anomalies are most certainly coming from contributions to k by the Kevlar strings as well as the Manganin wires. The fit through the curve is a 5th order polynomial. For this experiment one can assume to first order that the magnetic field and temperature dependence of the thermal link are independent from each other insofar as $k(H, T)$ can be expressed as a product of the form $k(H, T) = k_H(H)k_T(T)$. As indicated above, the field dependence is fitted by a

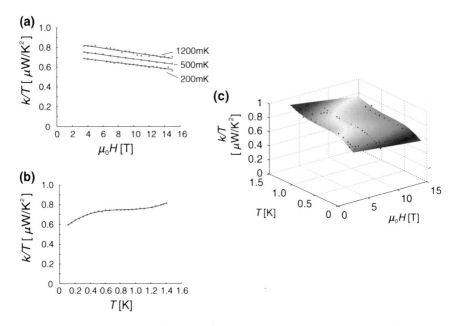

Fig. 4.21 a In *blue* is shown the thermal link k divided by temperature T as a function of magnetic field H for three different setup temperatures as indicated. The *red* curves are linear fits through the data. **b** The same quantity but this time as a function of temperature at an applied magnetic field of 8 T. The fit in *red* is a polynomial of fifth order. **c** The fit surface for the thermal link k divided by temperature T as a function of both T and magnetic field H. Also shown as *black dots* are the data points used to create the surface

linear polynomial and the temperature dependence by a 5th order polynomial. Besides the traces already presented, two further temperature dependences were measured at magnetic fields of 4 and 14 T.

Figure 4.21c shows the resulting surface together with the data points used to generate the fit. The standard deviation of the relative error of the measured data points k_{exp}/T from the surface k_{fit}/T, that is the standard deviation of the quantity $(k_{\text{exp}} - k_{\text{fit}})/k_{\text{fit}}$ is found to be $\approx 1\%$. It is this fitted surface for the thermal link that was used in the data analysis presented in the following chapter.

4.5.3 Specific Heat

Finally, an important last part of this chapter is the discussion of the parameters chosen for the AC specific heat measurements presented in this work. As discussed in Sect. 3.2.1 the most important parameter is the choice of excitation frequency. It is unusual in that it is of the order of several tens of mHz contrary to a more common value of above 100 Hz. This extremely low frequency is a consequence of the experiment being optimised for the magnetocaloric sweeps.

4.5 Characteristics and Details of Measurements on $Sr_3Ru_2O_7$

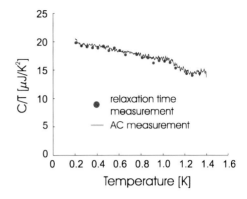

Fig. 4.22 Here the heat capacity C of sample C698J and addenda divided by temperature T as a function of T is plotted. The magnetic field at the sample is 8 T. In *red* the results of the relaxation time method are given whereas in blue the results of the AC specific heat method are shown

The frequency dependence of the AC specific heat measurement has been carried out at several temperatures and magnetic fields in the area of interest of the phase diagram. Overall the optimal frequency for the experimental conditions has been found to be 35 mHz. For this value the variation of the correction factor over the phase diagram is of the order of 1%.

Without wanting to discuss the physical meaning here, I show in Fig. 4.22 the temperature dependence of the heat capacity C of sample C698J and addenda divided by temperature T as measured by the relaxation time method (red) and AC technique (blue) with the above mentioned excitation frequency of 35 mHz. The magnetic field applied to the sample is 8 T.

At the position of the relaxation time measurement points the AC method is found to overestimate the specific heat on average by a factor of 1.01(1). Since this is consistent with 1 and an error of 1% is at the limit of achievable accuracy in this experiment, it was decided not to apply a correction factor to the AC specific heat results.

References

1. Pobell F (1992) Matter and methods at low temperatures, pp 105–137. Springer, Heidelberg
2. Lounasmaa OV (1974) Experimental principles and methods below 1K. Academic Press, London
3. Pobell F (1992) Matter and methods at low temperatures, pp 44–48. Springer, Heidelberg
4. du Chatenier FJ, Miedema AR (1965) Heat capacity below 1 K: observation of the linear term and the HFS contribution in some dilute alloys. In: Proceedings of the 9th international conference on low temperature physics, p 1029
5. Schink HJ, Löhneysen Hv (1981) Specific-heat of Apiezon-N grease at very low-temperatures. Cryogenics 21(10):591–592
6. Siqueira ML, Rapp RE (1991) Specific heat of an epoxi resin below 1 K. Rev Sci Instrum 62(10):2499–2500
7. Pobell F (1992) Matter and methods at low temperatures, pp 40. Springer, Heidelberg
8. Volokitin YE, Thiel RC, de Jongh LJ (1994) Heat capacity of thick film resistor thermometers and pure RuO_2 at low temperatures. Cryogenics 34:771

9. Ho JC, O'Neal HR, Phillips NE (1963) Low temperature heat capacities of Constantan and Manganin. Rev Sci Instrum 34:782
10. Ventura G, Barucci M, Gottardi E, Peroni I (2000) Low temperature thermal conductivity of Kevlar. Cryogenics 40(7):489–491
11. Peroni I, Gottardi E, Peruzzi A, Ponti G, Ventura G (1999) Thermal conductivity of Manganin below 1 K. Nuclear Physics B-Proceedings Supplements 78:573–575, 6th international conference on advanced technology and particle physics, Villa Olmo, Italy, OCT 05–09, 1998
12. Davey G, Mendelssohn K, Sharma JKN (1965) In: Proceedings of the ninth international conference on low temperature physics, p 1196
13. Goodrich RG, Hall D, Palm E, Murphy T (1998) Magnetoresistance below 1 K and temperature cycling of ruthenium oxide-bismuth ruthenate cryogenic thermometers. Cryogenics 38(2):221–225
14. Mackenzie AP, Haselwimmer RKW, Tyler AW, Lonzarich GG, Mori Y, Nishizaki S, Maeno Y (1998) Extremely strong dependence of superconductivity on disorder in Sr_2RuO_4. Phys Rev Lett 80(1):161–164
15. Mackenzie AP, Ikeda S, Maeno Y, Fujita T, Julian SR, Lonzarich GG (1998) Fermi surface topography of Sr_2RuO_4. J Phys Soc Jpn 67(2):385–388
16. Nishizaki S, Maeno Y, Mao ZQ (2000) Changes in the superconducting state of Sr_2RuO_4 under magnetic fields probed by specific heat. J Phys Soc Jpn 69(2):572–578
17. Mackenzie AP, Julian SR, Diver AJ, McMullan GJ, Ray MP, Lonzarich GG, Maeno Y, Nishizaki S, Fujita T (1996) Quantum oscillations in the layered perovskite superconductor Sr_2RuO_4. Phys Rev Lett 76(20):3786–3789
18. Kikugawa N, Mackenzie AP. Private communication
19. Mercure J-F (2008) The de Haas-van Alphen effect near a quantum critical end point in $Sr_3Ru_2O_7$. Ph.D. thesis. September

Chapter 5
Experimental Results and Discussion

The aim of this study is to investigate the entropy of $Sr_3Ru_2O_7$ in the vicinity of the proposed quantum critical end point. Of particular interest are the entropic properties of the material in relation to the phase formation of the anomalous 'electron nematic' [1]. The phase diagram as established before this project was discussed in detail in Sect. 2.2.2. The most important aspects are summarized in the following, in order to be able to put the detailed results presented here in context with the wider phase diagram. For this purpose Fig. 2.15 from Chap. 2 is reproduced below. For more details please refer to Sect. 2.2.2.

In Fig. 5.1a [2] the two first order metamagnetic sheets (green) are shown as a function of temperature, magnitude and direction of the applied magnetic field. The data were obtained from analysis of the out of phase component of the AC magnetic susceptibility. The lines of critical endpoints of these first order sheets are marked in black. Furthermore, the regions enclosed in the blue domes represent the volumes of phase space in which a 'nematic-like' signature has been observed in the resistivity. The main part of this chapter is concerned with the entropy as a function of temperature with a magnetic field applied parallel to the c-axis. In the diagram presented in Fig. 5.1a, this corresponds to an angle of 90°. In Appendix B results will be presented of an initial magnetocaloric study with the magnetic field applied at an angle of approximately 70° as well as with the magnetic field perpendicular to the c-axis.

In Fig. 5.1b a more detailed view of previous results for the magnetic field parallel to the crystallographic c-axis is given. Here the phase boundaries of the anomalous 'nematic' region as seen by magnetisation, AC magnetic susceptibilities and thermal expansion are shown [3, 4]. In particular the first order phase transition lines identified by the out of phase component of the AC magnetic susceptibility are indicated in green. Along the blue line anomalies in thermodynamic and transport measurements were observed. These were suggestive of a second order phase transition, but no proof of that exists in the literature. This line of anomalies will be referred to in the following as the 'roof' feature. Furthermore,

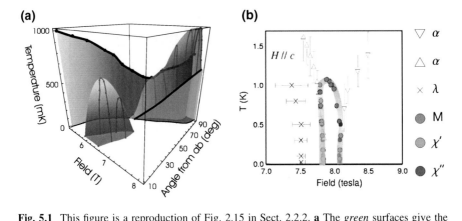

Fig. 5.1 This figure is a reproduction of Fig. 2.15 in Sect. 2.2.2. **a** The *green* surfaces give the position of the metamagnetic first order sheets in the phase diagram of $Sr_3Ru_2O_7$ as a function of temperature, magnitude of magnetic field and its direction as a function of angle from the crystallographic *ab* plane [2]. In *black* the line of critical endpoints is shown. In the regions under the *blue surfaces* 'electron nematic' behaviour has been observed [1]. **b** The phase diagram of $Sr_3Ru_2O_7$ as a function of temperature and magnetic field with the field applied parallel to the *c*-axis is shown. In particular the position of the phase transition lines (*green*) and thermodynamic anomalies (*blue*) enclosing the anomalous phase as seen by magnetisation M and AC magnetic susceptibility χ are presented. Furthermore, the position of thermodynamic crossovers as measured by linear magnetostriction λ and thermal expansion α are given [3, 4]

the positions of thermodynamic crossovers are indicated in the phase diagram (white triangles and crosses) based on magnetostriction, thermal expansion and AC susceptibility measurements.

5.1 Caloric Studies of Magnetic Phase Transitions in $Sr_3Ru_2O_7$

In this section I present the results of caloric studies of the magnetic phase transitions observed in $Sr_3Ru_2O_7$ with the magnetic field applied parallel to the crystallographic *c*-axis.

5.1.1 Evolution of Entropy across Phase Transitions as a Function of Field

5.1.1.1 Magnetic Field Sweeps at 250 mK

In order to investigate the entropy in the important temperature-field region of the phase diagram of $Sr_3Ru_2O_7$, both specific heat and magnetocaloric measurements were performed. In Fig. 5.2a, the result of a single AC specific heat measurement

Fig. 5.2 **a** An isothermal sweep of the specific heat c divided by temperature T as a function of magnetic field H. The temperature of the sample is $T_S = 250$ mK. Further labelled are the positions of the two first order transitions (*1*) and (*2*) as well as the metamagnetic crossover (*3*). In the *inset* the position of the sweep measurement in the phase diagram is shown in *blue*. **b** The temperature of the sample during a sweep of the magnetic field at a rate of 0.02 T/min (*blue*) and − 0.02 T/min (*red*) as a function of magnetic field. (*1*), (*2*) and (*3*) again mark the position of the first order transition lines and the metamagnetic crossover, respectively

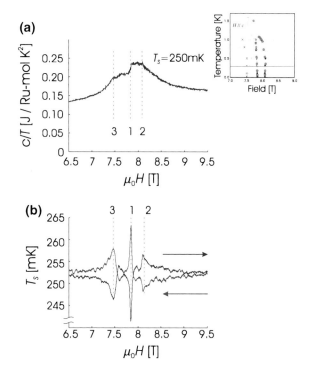

as a function of magnetic field is shown for a sample temperature of $T_S = 250$ mK.[1] Here one can clearly identify the two features (1) and (2) expected to be first order phase transitions as well as the low field crossover (3) as points of abrupt changes in the field dependence of the specific heat c (the details of which will be discussed later on). Transition (1) in particular shows a clear sharp rise in c within the intrinsic width of the transition of the order of 50 mT. This transition width is consistent with other experiments such as, for example, magnetic susceptibility [3] and is most probably due to impurity effects. However, it is not possible to infer anything about the change of entropy, i.e. latent heat, associated with these transitions from the specific heat measurement.

In order to investigate the transitions further, magnetocaloric measurements were performed. Figure 5.2b shows the temperature of the sample as a function of magnetic field while sweeping the same field at a rate of 0.02 T/min (blue) and −0.02 T/min (red).[2] The temperature change of the sample during an up and down

---

[1] The measurement was performed with a temperature modulation at a frequency of 70 mHz and an oscillation amplitude not exceeding ±3% of the sample temperature. The sweep rate was 0.01 T/min in order to keep the temperature variation due to the magnetocaloric effect to below ±1.5%. The data were analysed as discussed in Sects. 3.2.1 and 4.4.2.

[2] The magnetic fields for all magnetocaloric measurements have been corrected for a known hysteresis in the field sweep of ±15 mT.

sweep is symmetric about an average temperature of approximately 250 mK, as expected for the magnetocaloric effect in the strongly non-adiabatic limit (see Sect. 3.2.2).

Following the blue curve from low fields one can clearly identify the two transitions (1) and (2) as well as the metamagnetic crossover feature (3) at lower field. At transition (1), upon entering the anomalous phase from lower fields, an abrupt cooling of the sample to below the average temperature is observed. Equally, one can see that the sample temperature increases above the average temperature at transition (2). From this it is already possible to conclude that the sample has an increased entropy inside the phase compared to the surrounding normal states. The quantitative analysis will be given further on.

This result is in particular remarkable for transition (1) since here the magnetisation of the sample is increasing with increasing magnetic field. Intuitively an increase in magnetisation is associated with magnetic order, i.e. a lowering of the entropy. This suggests a significant change in the internal degrees of freedom when entering the anomalous 'nematic' phase region.

5.1.1.2 Clausius–Clapeyron Relation

Counterintuitive as the result may seem, it is consistent with the well known Clausius−Clapeyron relation. As discussed in Sect. 3.1.2, this law relates the curvature of the phase transition line in the $H - T$ diagram, i.e. dH_C/dT_C where H_C is the transition field and T_C is the transition temperature, with the change in entropy ΔS and magnetisation ΔM across the transition:

$$\frac{dH_C}{dT_C} = -\frac{\Delta S}{\Delta M}. \tag{5.1}$$

Magnetisation measurements have shown that M increases monotonically with magnetic field (see for example Fig. 2.13 in Sect 2.2.2). Furthermore dH_C/dT_C is negative/positive for the first order transition (1)/(2). A sign analysis of the Clausius–Clapeyron relation would therefore qualitatively predict an increase of the entropy ΔS across transition (1), which is in agreement with the measurements presented. Similarly, it holds that across the second first order transition (2) a decrease of the entropy would be predicted due to the difference in curvature of dH_C/dT_C, again in qualitative agreement with the temperature change of the sample at the transition during a field sweep.

An important aspect of the Clausius–Clapeyron relation is its dependence on only two underlying assumptions, namely that it relates to (i) first order phase transitions between (ii) two thermal equilibrium phases (see Sect. 3.1.2). A quantitative agreement of the data with this relation is therefore a necessary condition to show that the observed transitions are true equilibrium transitions. In order to verify this it is necessary to calculate the change of entropy $S(H) - S(H_0)$ as a function of magnetic field H relative to a reference field H_0 from the magnetocaloric sweeps and specific heat measurements.

5.1 Caloric Studies of Magnetic Phase Transitions in Sr$_3$Ru$_2$O$_7$

As shown in Sect. 3.2.2, this is given by

$$S(H) - S(H_0) = -\int_{H_0}^{H} dH \left\{ \frac{k}{T_S} \frac{T_S - T_{\text{bath}}}{\dot{H}} + \frac{C(T_S, H)}{T_S} \frac{dT_S}{dH} \right\}. \quad (5.2)$$

Here k is the thermal conductivity of the heat link between the sample platform and the copper ring, as discussed in Sect. 4.2. T_S is the measured sample temperature during a sweep, T_{bath} is the temperature of the copper ring as measured, C is the heat capacity of the setup (including addenda) and $\dot{H}$ the sweep rate.

Therefore, in order to establish the entropy jump along the phase transition line, one has to first carry out the magnetocaloric sweep over a range of temperatures and secondly to know the heat capacity C of the setup over the important range of the phase diagram. Figure 5.3a shows the temperature change of the sample as a function of magnetic field for a range of thermal bath temperatures. All curves have the same scale as shown but were offset from each other for clarity of presentation. The offset is in proportion to their temperature difference. Figure 5.3b shows the heat capacity of the setup divided by temperature as a function of magnetic field. The curves again are offset for clarity but have the same scale. The data are dominated by the heat capacity of the sample with a small contribution from the addenda. These measurements were used to create an interpolation surface in order to be able to approximate the heat capacity of the setup over the whole temperature-field regime.

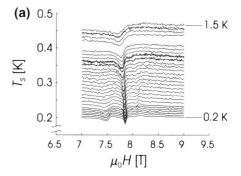

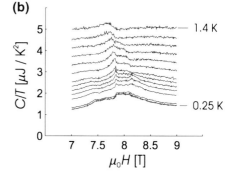

Fig. 5.3 a The temperature of the sample T_S during the sweeping of the magnetic field is shown as a function of magnetic field H. The temperature scale corresponds to the lowest trace at 200 mK whereas all other traces are offset in proportion to the difference of the thermal bath temperature during the sweep. b The AC heat capacity C of the setup including a small addenda contribution divided by temperature T as a function of magnetic field H for several sample temperatures. The specific heat scale corresponds to the lowest temperature trace of 0.25 K. All other traces have the same scale but are offset in proportion to their temperature difference for clarity

Based on these measurements and using equation 5.2 one can calculate the field derivative of the entropy dS/dH, and from there, the relative change of the entropy ΔS with respect to a reference field $S(H_0)$, i.e. $\Delta S(H) = S(H) - S(H_0)$. Furthermore, since in Sect. 4.4.2 it was established that the entropy change of the addenda as a function of field is negligible in the temperature-field regime studied here, it follows that the calculated entropy changes are those of the sample.

The resulting field derivatives of the entropy as a function of magnetic field are shown in Fig. 5.4a. The curves have been offset for clarity (please note the inverted offset direction compared to previous graphs). These curves look very similar in appearance to the magnetocaloric sweeps themselves. The reason for this is that the integrand in (5.2) is dominated by the first term which is proportional to the temperature deviation of the sample from the thermal bath. The maxima and minima in these curves at the transition correspond to the fastest changes in entropy. These extrema can therefore be used to extract the position of the phase transitions in the $H - T$ plane.

In Fig. 5.4b the resulting changes in entropy with respect to 4 T, $\Delta S = S(H) - S(4T)$, divided by temperature T are shown as a function of magnetic field. The transitions (1) and (2) are clearly identifiable as sharp features in the entropy. The metamagnetic crossover (3) on the other hand is a relatively smooth feature in the entropy. In particular for transition (1) the jump in entropy is clearly visible and it will be this transition that is analysed in more detail in the following discussion.

Figure 5.5a shows the entropy as a function of magnetic field around transition (1) for $T = 450$ mK (blue curve). In red are shown a third order polynomial fit to

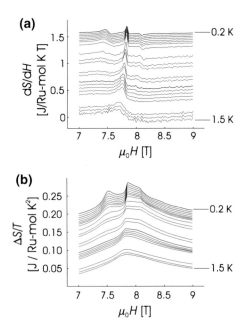

Fig. 5.4 In **a** the derivative of entropy S with respect to magnetic field H is given as a function of magnetic field. The curves are reconstructed from the previous data. The scale on the left corresponds to the highest temperature trace at 1.5 K. The other traces have the same scale but are offset in proportion to their difference in sample temperature for clarity. In **b** the change in entropy S divided by sample temperature T is shown as a function of magnetic field H. The entropy scale corresponds to the trace at 1.5 K. All other traces have the same scale but are offset in proportion of the difference in sample temperature as before

5.1 Caloric Studies of Magnetic Phase Transitions in $Sr_3Ru_2O_7$

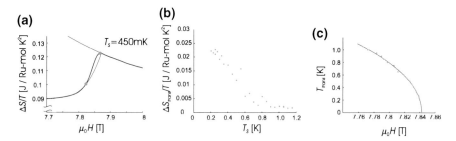

Fig. 5.5 **a** shows the entropy change ΔS divided by temperature T as a function of magnetic field H (*blue curve*) across the main first order metamagnetic transition marked as (1) in Fig. 5.2. The sample temperature is $T_S = 450$ mK. In *red* are shown polynomial fits to the entropy of the thermodynamic phases below and above the transition. The *green points* are marking the range over which the transition occurs determined by a deviation of the entropy from the polynomial fits as described in the text. The difference in entropy between these two points is giving ΔS_{trans}. This quantity divided by temperature is shown in **b** for a range of sample temperatures T_S. Furthermore in **c** the position of the first order transition in the $H-T$ diagram is shown (*blue*) together with a power law fit (*red*) through the data as explained in the text

the low field entropy and second order polynomial fit to the high field entropy. The magnetic field range for the phase transition is determined by the points where the entropy starts deviating from the polynomials significantly. The cut-off value here has been chosen to be a deviation above 0.4 mJ/Ru-mol K^2 which is approximately three times the standard deviation of the data from the polynomial fit. These points of deviation are marked in green in the plot. The difference in entropy between them is taken as the jump in entropy across the transition, whereas the difference in magnetic field as the width of the transition at this temperature. Figure 5.5b shows the entropy jump divided by temperature as determined by the above procedure as a function of bath temperature. As one can see, the entropy jump decreases continuously from its maximum value of 22.5 mJ/Ru-mol K^2 at low temperatures to zero at around 800 mK. This is consistent with previous measurements that traced the out-of-phase component in the AC magnetic susceptibility as an indication of a first order transition to around the same temperature [3]. The remaining 'jump' in entropy at temperatures above 0.9 K is an artefact of the still rapid change in entropy in combination with the 'blind' application of the procedure described above to extract the jump.

In order to compare this jump in entropy to the jump one would expect from the change in magnetisation across the transition according to the Clausius–Clapeyron equation one needs the quantity dH_C/dT_C, i.e. how the critical field of the transition changes as a function of temperature. For this the midpoint of the transition is extracted from the data as the point of fastest change in entropy as a function of field. The result of this is shown in blue in Fig. 5.5c. Furthermore, in red a power law fit of the form

$$H_C = aT_C^b + c \tag{5.3}$$

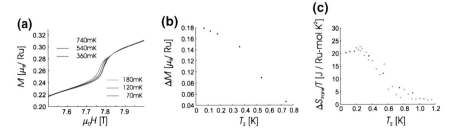

Fig. 5.6 a The magnetisation M of $Sr_3Ru_2O_7$ as a function of magnetic field H across the main metamagnetic first order transition for several temperatures as indicated by the colour scheme. The data has kindly been provided by R.S. Perry. **b** The extracted jump in magnetisation ΔM across the transition for the sample temperatures T_S (see details given in text). Finally **c** is giving in *red* the entropy jump ΔS_{trans} across the first order transition divided by sample temperature T as calculated by the Clausius–Clapeyron equation from the magnetisation jump ΔM shown in the previous figure. In *blue* is given for comparison the entropy change across the transition as already shown in Fig. 5.5b

is given, with a, b and c being free parameters. The fit results in $a = -0.069(2)$ T/K, $b = 2.1(2)$ and $c = 7.842(2)$ T. It is this fit that is used to evaluate dH_C/dT_C at appropriate temperatures.[3]

The final quantity needed is the jump in magnetisation across the transition. Here data for the magnetisation as a function of magnetic field across the transition were kindly provided by R.S. Perry. This is shown in Fig. 5.6a. In order to be consistent with the previous analysis, the midpoint of the transition at the given temperature was first established as the point of fastest rise in magnetisation.[4] Then the change in magnetisation over a width identical to the width extracted from the entropy change at the appropriate temperatures was measured. The resulting magnetisation jumps as a function of the transition temperature are shown in Fig. 5.6b. From these and the quantity $dH_C(T)/dT$ it is now possible to calculate the entropy jump one would expect according to the Clausius–Clapeyron relation. These are shown in red in Fig. 5.6c. The overall agreement with the entropy jump directly measured across the transition as shown in blue is very good within the accuracy of the presented analysis.

It can therefore be said in summary that the results presented here are in good qualitative as well as quantitative agreement with the Clausius–Clapeyron relation. This not only confirms that the entropy of the anomalous phase is higher than that of the surrounding 'normal' phases, it also shows that the entropic data of the

---

[3] Though a power law fit over a limited region with three free parameters cannot be considered very reliable it is worth pointing out that according to the Clausius–Clapeyron relation the fact that the fit of the power b is consistent with 2 results in the entropy jump divided by temperature, i.e. $\Delta S/T$, to be proportional to the jump in magnetisation across the transition, ΔM.

[4] These were found to be approximately 50 mT lower than measured in the magnetocaloric sweeps. This is an indication that during the magnetisation measurements by R.S. Perry the magnetic field was applied at a small angle with the c-axis.

5.1 Caloric Studies of Magnetic Phase Transitions in Sr$_3$Ru$_2$O$_7$

observed transitions are quantitatively in agreement with that expected from equilibrium first order thermodynamic phase transitions.

5.1.2 Specific Heat Signature of Phase Transitions

In the previous section the changes in entropy as a function of magnetic field were studied in detail primarily with the magnetocaloric effect. This allowed in particular an investigation of the two first order transition line boundaries of the anomalous phase. However, in order to investigate the 'roof' feature and in particular to discuss its thermodynamic nature a specific heat study as a function of temperature is more suitable.

In Fig. 5.7a, traces of AC specific heat measurements at four different fields are shown as a function of sample temperature T_S.[5] The position of these

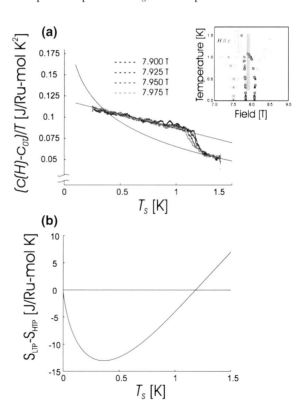

Fig. 5.7 a The specific heat $c(H)$ minus the specific heat at zero field $c_{0\,T}$ divided by temperature T as a function of temperature for four magnetic fields. The position of these measurements in the phase diagram is indicated by the *blue line* in the *inset*. The *continuous red curve* in **a** is a linear fit to the data in the low temperature phase (LTP) and the continuous blue curve is the extrapolation of the high temperature phase (HTP) specific heat divided by temperature. **b** The difference in entropy $S_{\text{LTP}} - S_{\text{HTP}}$ between these two phases as a function of temperature based on the fits shown in **a**

---

[5] The same temperature modulation as discussed in the previous section has been employed. Furthermore the magnetic field was kept constant and the temperature of the copper ring was continuously lowered at a rate of 2.25 mK/min.

measurements in the phase diagram is indicated in the inset. To be more precise, the difference between the specific heat $c(H, T_S)$ at field H and temperature T_S and the zero field specific heat $c(0, T_S)$ divided by temperature T is plotted. Since $Sr_3Ru_2O_7$ is a Fermi liquid in zero field, this differs from $c(H, T_S)/T_S$ only by a constant offset. The continuous red and blue lines are fits to, respectively, the high temperature and low temperature specific heat data as discussed later on.

At the position of the 'roof' a sharp rise in specific heat is observed upon lowering the temperature. Furthermore, the feature disperses with field as expected, i.e. it moves to lower temperatures as the magnetic field is increased. The width in temperature for the transition at 7.9 T is approximately 75 mK. In Sect. 4.4.3 it was shown that the experimental setup is capable of resolving the superconducting transition in Sr_2RuO_4 to within 45 mK. It therefore has to be concluded that at least some of the width of 75 mK of the feature observed here is due to the properties of the sample rather than of the instrument.

Intriguingly, the specific heat divided by temperature of the anomalous phase, c/T, keeps rising approximately linearly to even lower temperatures. An important question is whether this linear rise extends to lower temperatures. Since at the observed phase transition the entropy of the high temperature phase is the same as that of the low temperature phase the following equality has to hold,

$$\int_0^{T_C} \left\{\frac{c_{LTP}}{T}\right\} dT = \int_0^{T_C} \left\{\frac{c_{HTP}}{T}\right\} dT, \tag{5.4}$$

with T_C the transition temperature, c_{LTP} the specific heat of the low temperature phase and c_{HTP} the theoretical specific heat of the high temperature state. This can be rewritten as

$$\int_0^{T_C} \left\{\frac{c_{LTP} - c_{HTP}}{T}\right\} dT = 0. \tag{5.5}$$

This therefore gives a possibility of checking the consistency of the assumption of the mathematical form of the low temperature phase specific heat provided one knows sufficiently well the theoretical behaviour of the specific heat of the high temperature phase. As has been shown previously by Perry et al. [5], close to the critical field the specific heat divided by temperature, c/T, has a singular contribution that is logarithmically diverging to lower temperatures. For this project, PPMS measurements were repeated with a small single crystal at the exact field of 7.9 T.[6] The measurements were performed between 0.5 mK and 30 K. The singular part of the high field state between 1.2 and 5 K is found to be well described by

---

[6] The weight of the $Sr_3Ru_2O_7$ sample used is 2.4 mg and was cut from a piece directly adjacent to the single crystal used primarily in this work.

$$\frac{c_{\text{sing}}(T)}{T} = -a\log(T/1K), \qquad (5.6)$$

with $a = -0.42(5)$ J/Ru-mol K^2. In Fig. 5.7a this singular part plus an offset for the normal Fermi liquid contribution is shown as the blue curve. The red curve furthermore is a linear fit to the specific heat of the low temperature state. One therefore has all necessary quantities to evaluate the left hand side of equation 5.5. The resulting curve is shown in Fig. 5.7b. This entropy difference curve crosses zero at approximately 1.2 K. Based on the assumptions that (a) the high temperature phase specific heat has a logarithmic singularity at zero temperature and (b) that the 'roof' feature is a true second order phase transition, one can conclude that the data are consistent with the low temperature state specific heat divided by temperature, c_{LTP}/T, to rise linearly to zero temperature.

Finally, the specific heat as function of temperature was measured for several magnetic fields close to or at the two first order boundaries of the anomalous phase. Figure 5.8a shows three traces of $(c(H, T) - c(0, T))/T$ as a function of temperature for 7.83, 7.84 and 7.8475 T. Coming from high temperatures, the anomaly upon entering the anomalous phase at 1.2 K is clearly visible [indicated by (A)]. Then towards lower temperatures, between 0.4 K and 0.7 K, the first order transition is crossed upon going out of the anomalous phase. It appears relatively wide in temperature. This is due to the combination of the intrinsic width

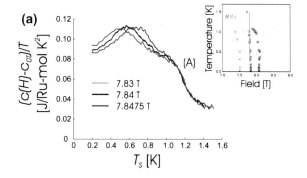
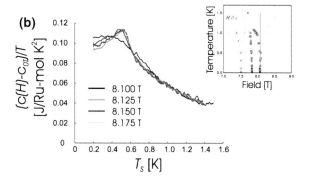

Fig. 5.8 The difference between the specific heat c at magnetic field H, $c(H)$, and the zero field specific heat $c_{0\,T}$ as a function of temperature for the magnetic fields indicated in the graphs. The *insets* in **a** and **b** are indicating the position of the measurements in the phase diagram via the two blue lines. The letter (A) in **a** is marking the position of the 'roof' feature as discussed in the previous figure

in magnetic field of 50 mT of the transition and the steep slope of the transition line in the $H - T$ diagram.

Figure 5.8b on the other hand shows $(c(H, T) - c(0, T))/T$ for magnetic fields close to the second first-order transition. Since the slope of the transition line in the $H - T$ diagram is even steeper than for the transition previously discussed, it is not possible to resolve the features coming from the 'roof' and the transition line.

In summary, it was shown in this section that the 'roof' signature as observed by other experiments is also reflected in specific heat measurements as an anomaly, which is consistent with a second order phase transition for the magnetic field applied along the crystallographic c-axis. If one furthermore assumes that the high temperature phase specific heat has a logarithmic singularity at zero temperature, than the data are consistent with the specific heat of the anomalous low temperature phase divided by temperature, c_{LTP}/T, to be linearly increasing towards zero.

5.1.3 Discussion

In this section I have given the results of a detailed investigation of the anomalous phase and the surrounding phase transitions. In the following I will summarize the most important results and discuss their relevance and relation to other experiments.

5.1.3.1 Phase Transition Characterisation

In the first part of the chapter, I showed the results of a detailed entropy study of the proposed thermodynamic transitions in $Sr_3Ru_2O_7$ via a series of magnetocaloric experiments. In particular the features surrounding the new anomalous phase region for the magnetic field applied parallel to the c-axis were investigated. In Fig. 5.9 the results of this study are compared to the phase diagram established in previous experiments. In **a** Fig. 2.15b is reproduced. **b** on the other hand shows a colour contour plot of the derivative of the entropy S with respect to magnetic field H, $\partial S/\partial H$, divided by temperature T, in the temperature field region studied in the work presented here. The smoothed surface has been reconstructed from the data obtained in the magnetocaloric sweeps presented in this section.

Finally, in Fig. 5.9c the graphical overlay of both with no further adjustments to the data is shown. The features in the change of magnetic entropy are coincident with the previously established signatures in the phase diagram. The two first order transitions correspond to sharp positive (red end of the colour scale) and negative (blue end of the colour scale) peaks in $\partial S/\partial H$. This is consistent with the interpretation of latent heat associated with these features. However, it is also clear that these proposed first order transitions are not infinitely sharp but have a finite width

5.1 Caloric Studies of Magnetic Phase Transitions in Sr$_3$Ru$_2$O$_7$

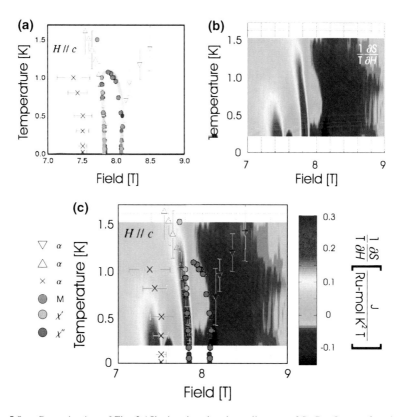

Fig. 5.9 **a** Reproduction of Fig. 2.15b showing the phase diagram of Sr$_3$Ru$_2$O$_7$ as a function of temperature and magnetic field based on previous measurements of magnetisation M, complex AC magnetic susceptibility χ, linear magnetostriction λ and thermal expansion α [3, 4]. *Green lines* represent proposed first order transitions whereas the *blue line* represents the 'roof' feature, proposed to be a second order phase transition. **b** The colour scale represents the partial derivative of entropy S with respect to magnetic field H, $\partial S/\partial H$, divided by temperature T over the same temperature-field region as **a**. Finally in **c** the two graphics are overlaid in accordance to their field and temperature axes without further adjustment to the data

of the order of 50 mT, consistent with previous data from, for example, AC magnetic susceptibility.

The signature of the 'roof' feature in the magnetocaloric data is a sudden change in the field derivative of the entropy. In particular the anomalous phase is characterised in this data by a significantly smaller magnitude of $\partial S/\partial H$ than the surrounding 'normal' states. This is equivalent to saying that the field dependence of the entropy in the anomalous phase region is weaker than in the surrounding region in phase space. According to the Maxwell relation 3.9 (see Sect. 3.1.1)

$$\left.\frac{\partial S}{\partial H}\right|_T = \left.\frac{\partial M}{\partial T}\right|_H \quad (5.7)$$

the entropy result is consistent with the previously reported reduced temperature dependence of the magnetisation M inside the anomalous phase region [3], see also inset in Fig. 2.12b, Sect. 2.2.2).

Finally, I would like to remark that the crossover feature at 7.5 T is at low temperatures coincident with a sign change in $\partial S/\partial H$, which is equivalent to a maximum in entropy S. At higher temperature the exact position of the crossover is less easily defined due to difficulties in separating background effects from the feature.

In the following, I will in detail address several questions regarding this phase diagram. First I will discuss whether the observed features bounding the anomalous phase in magnetic field are indeed first order. Secondly the nature of the 'roof' feature will be investigated. Finally, I will discuss the consequences of the observed specific heat for theoretical models concerning the novel phase.

First Order Phase Transitions

Though widely assumed, it is not a priori clear that the anomalous phase region is indeed bound by first order phase transitions as a function of magnetic field. Indeed, after the initial observation of the novel quantum phase by Grigera et al. [3], it was proposed by Binz et al. [6] that the anomalous region is a mixed state of domains of low magnetisation and high magnetisation phases in an extension of the concept of Condon domains [7]. In other words, the region would not be a thermodynamic equilibrium phase. It was shown that even in this scenario the phase would be bounded as a function of magnetic field by superlinear rises in the magnetisation. In this theoretical model the width in magnetic field of the anomalous phase region as well as the size of the magnetisation jumps depend on the demagnetization factor n of the sample and therefore ultimately on sample shape. However, previous studies did not find such a significant shape dependence making this particular theoretical proposal unlikely to be realised in $Sr_3Ru_2O_7$. Nevertheless, it highlighted the possibility of non-equilibrium phenomena in the magnetic phase diagram of $Sr_3Ru_2O_7$.

Two further characteristics of the transition features motivate a further experimental investigation. First, they are not sharp but have a finite width in magnetic field of the order of 50 mT. Small variations in this width are observable for different samples. This finite width could be caused by several effects. Most likely among them is a broadening of the transitions due to a finite impurity content in these ultra clean samples. This is supported by the observation that the phase, which has a width of approximately 0.3 T, is not observable as a well defined region in samples that have an in-plane residual resistivity that is still low (3 μΩ cm) but an order of magnitude higher than in the ultraclean samples studied here $(\rho_0 \leq 0.5\,\mu\Omega\,cm)$ [8]. This indicates that the anomalous phase and its bounding phase transitions are highly sensitive to impurities. A second possibility is a distribution of transition temperatures in the sample due to demagnetisation effects. This effect, though related to the demagnetisation factor n, is fundamentally different from the above discussed formation of Condon domains across the whole anomalous phase region.

Secondly, even though the transitions are metamagnetic in nature there is no true divergence in the AC magnetic susceptibility at the proposed critical endpoints [9, 10] as one would expect in the thermodynamic limit and as has been observed in other metamagnetic materials [11]. One possible reason for this is the above discussed finite width of the transition. A second possibility is that the metamagnetic transition is not occurring at $q = 0$ but at a small but finite q. In this case the jump one observes in the $q = 0$ longitudinal magnetisation is a reflection of the finite q singularity [10]. A possibility for such a scenario is the formation of a spin-spiral state which has a transverse magnetisation component as discussed by Berridge et al. [2]. Together, the theoretical possibility of non-equilibrium phenomena and the unusual experimental observations made a further investigation of the metamagnetic features necessary.

In the work presented in this thesis I set out to investigate whether the experimentally observed change in entropy ΔS and in magnetisation ΔM across the width of the proposed first order transition features as well as the curvature of the transition line in the temperature-field phase diagram are qualitatively and quantitatively consistent with the Clausius–Clapeyron relation. As discussed in Sect. 3.1.2 the only two assumptions that have to be fulfilled in order for this relation to hold are that the observed features are (a) first order phase transitions between (b) equilibrium states of the material. If any of these two assumptions do not hold then the Clausius–Clapeyron relation does not have to be followed.

In the data presented here it is found that both transitions bounding the anomalous phase as a function of magnetic field are qualitatively consistent with this relation, i.e. that the sign of the curvature of these transition lines is consistent with the observed changes in entropy and magnetisation. A quantitative test was only possible for the first of the two transitions at 7.8 T which has a much stronger experimental signature. Here it was shown that the data are also quantitatively consistent with the Clausius–Clapeyron relation.

This result puts important constraints on theoretical proposals for the physics of $Sr_3Ru_2O_7$. First, any model assuming that the observed features are not true first order transitions between equilibrium states has to address the issue of why the Clausius–Clapeyron relation is observed to hold within experimental resolution. Secondly, if one assumes that $Sr_3Ru_2O_7$ is in thermal equilibrium across the whole phase diagram as studied in this thesis, then the direct consequence of the magnetocaloric measurements presented is that the entropy of the anomalous phase is higher between the first order transitions than that of the 'normal' states at lower and higher magnetic fields. This indicates that the novel quantum phase has additional degrees of freedom compared to the surrounding 'normal' states. In particular, it has to be pointed out, that the low field limit of the entropy jump divided by temperature, $\Delta S/T$, is of the order of 25 mJ/Ru-mol K^2, corresponding to 20% of the Sommerfeld coefficient at zero field of 110 mJ/Ru-mol K^2. Therefore, a significant part of the electronic degrees of freedom of $Sr_3Ru_2O_7$ has to be involved in the phase transition.[7] To my best knowledge, none of the proposed microscopic theoretical models for the physics of $Sr_3Ru_2O_7$ has reproduced the particular sign of the curvature of the phase transition lines and therefore the equivalent observation of a higher entropy inside the anomalous phase region as compared to the low and high field 'normal' states.

Finally, it should be remarked that the above discussion does not exclude the existence of domains in the anomalous phase. Particularly, if the domains are characterised by an order parameter that couples only weakly to the longitudinal magnetic field, it can be assumed that the effect of the existence of domains on the physics of the phase transitions does not alter their properties to first order and represents a weak perturbation. One such order parameter could for example be the transverse magnetisation. The entropy associated with domain walls can also be assumed to contribute a negligible part to the overall entropy of the system if the domains are of sufficient size. This is most likely the case in $Sr_3Ru_2O_7$ since Mercure et al. [12] were able to show that if domain walls are restricting the electron orbits at the centre of the quantum oscillation phenomena, than they should have an average separation of 0.5 μm within the plane. Each domain would therefore contain at least of the order of 10^6 unit cells.[8]

Characteristics of the 'Roof' Feature

In the discussion so far it has not been shown that the anomalous region in phase space is indeed a well defined novel quantum phase that is separated in the $H - T$ plane of the phase diagram by phase transitions from the surrounding 'normal' states. This would only be the case if the thermodynamic features at the 'roof' of the anomalous phase region are indeed due to a second order phase transition and not a crossover. I will therefore in the following discuss the extent to which the observed anomaly in specific heat is consistent with a second order phase transition.

At the observed transition at 7.9 T and 1.19(1) K the specific heat shows as a function of temperature a rapid increase of 25(5) mJ/Ru-mol K^2 over a range of approximately 75 mK. This is reasonably sharp to identify it as a thermodynamic feature. However, in Sect. 4.4.3 it was shown that the second order superconducting transition of Sr_2RuO_4 in zero field at 1.5 K can be resolved in the experimental setup under comparable conditions to within 45(5) mK. This indicates that the

---

[7] Another way of stating the significance of the jump of entropy ΔS is that at its maximum at around 400 mK its absolute value is of the order of 0.1% of $R\ln(2)$/Ru-mol, with R being the gas constant and the '2' a somewhat arbitrary choice representing the degrees of freedom of a high temperature system of fluctuating independent spins.

[8] The in-plane a and b lattice constant of the unit cell are approximately 0.4 nm. A domain is therefore expected to contain of the order of $(0.5 \ \mu m)^2/(0.4 \ nm)^2$ unit cells if one assumes that the domains are quasi-two-dimensional due to the quasi-two-dimensional character of the electronic system.

5.1 Caloric Studies of Magnetic Phase Transitions in Sr$_3$Ru$_2$O$_7$

observed width of 75 mK in Sr$_3$Ru$_2$O$_7$ is due at least in part to the properties of the sample.

There are several possible causes for this. First, the intrinsic second order phase transition can be broadened by impurities. If one assumes that the energy scale of the observed broadening of the first order transitions given by $\mu_B \Delta H = \mu_B \times 50$ mT is also the energy scale of broadening in temperature, than one would expect an intrinsic width of the order of $\Delta T = \mu_B \Delta H / k_B - 33$ mK. However, this purely energetic argument assumes that there is no dispersion of the phase transition temperature as a function of magnetic field. Indeed, as has been shown in Fig. 5.7, the midpoint of the feature in specific heat moves to lower temperatures as a function of field at a rate of the order of 0.8 K/T. This implies, that the overall width of the 'roof' feature includes contributions of broadening in both temperature and field. What is meant by this is most easily seen at the two first order transitions in Sr$_3$Ru$_2$O$_7$ (see for example Fig. 5.9). Here the transition temperature changes rapidly as a function of magnetic field. This leads to very broad features in the specific heat as a function of temperature due to the finite width of the transitions in magnetic field of the order of 50 mT. In the case of the 'roof' feature this would lead to a broadening of the order of 50 mT $\times$ 0.8 K/T = 40 mK. Of course, this effect and the energetic broadening of 33 mK are not expected to simply add. However, in total they should account for at least half of the observed width of the transition. A more detailed study of the transition in temperature and magnetic field is necessary for a more quantitative analysis.

The second possibility for the observed broadening relates to the physics of the novel quantum phase. If one assumes that for the magnetic field applied parallel to the crystallographic c-axis, the 'roof' represents a true second order phase transition, then a symmetry breaking thermodynamic order parameter will be associated with it. The transport measurements reported by Borzi et al. indicate that this symmetry can also be broken by a small magnetic field component in the crystallographic ab-plane. Such a component can be present in the experiment here due to a possible misalignment of the sample of the order of 1°. This situation would then be very similar to a ferromagnetic second order phase transition with a small but finite magnetic field applied to the sample. In this case, no true sharp second order phase transition is observed but a crossover from the paramagnetic to the ferromagnetic state. The width in temperature of this crossover is increasing with increasing field. In analogy, what one would expect for the 'roof' feature in Sr$_3$Ru$_2$O$_7$ in this scenario is that the second order phase transition turns into a crossover of finite width if the magnetic field is not applied perfectly parallel to the crystallographic c-axis. This proposed explanation for the experimental result predicts that if one increases the in-plane magnetic field component deliberately in an angular study, the width in temperature of the feature in specific will increase. This will have to be investigated in a future study.

The assumption that the observed specific heat feature is a broadened second order phase transition is also consistent with the magnetocaloric experiments presented in this thesis. They show that the field derivative of the entropy $\partial S / \partial H$ changes significantly at the 'roof' feature but no peak associated with latent heat is observed. Since entropy is a first derivative of the free energy, this is therefore an observation of an effective non-analyticity of a second order derivative of the free energy within the width of the observed feature with the first derivative being continuous over the same range. This is effectively the definition of a second order phase transition.

In summary, the above presented arguments show that even though the observed specific heat signature of the 'roof' has a small but finite width in temperature, it is nevertheless consistent with a true second order phase transition for the magnetic field applied parallel to the c-axis. The novel quantum phase in this scenario is a region in phase space that is distinct from the surrounding 'normal' states in that it is completely separated from them by first and second order phase transitions as a function of applied magnetic field in the case of the field oriented parallel to the c-axis.

Properties of the Novel Quantum Phase

The above conclusion of the 'roof' feature being a second order phase transition has the important implication that at the transition temperature the entropy of the high temperature phase is identical to that of the anomalous low temperature phase. The implication is that the integral of the specific heat of the low temperature phase from zero temperature up to the transition temperature has to be equal to the integral of the hypothetical specific heat of the high temperature phase over the same temperature region. In this section it was shown that the experimental data is consistent with this if one (a) assumes that the specific heat of the high temperature phase has a logarithmic singularity at zero temperature and that (b) the specific heat c of the low temperature phase as a function of temperature has the form

$$c = aT - bT^2 \tag{5.8}$$

with the quadratic correction b to the linear specific heat of standard Fermi liquid theory being $b = 33(1)$ mJ/Ru-mol K^3. This means that the value of the specific heat divided by temperature at 1 K is 33 mJ/Ru-mol K^2 smaller than the zero temperature Sommerfeld coefficient which is of the order of 230 mJ/Ru-mol K^2. This is on the one hand a significant deviation from standard Fermi liquid theory. On the other hand Mercure et al. [12] have conclusively proven the existence of quantum oscillations in the novel phase indicating the existence of fermionic Landau quasiparticles and the associated Fermi surface. In the following two possible explanations reconciling these experimental observation will be discussed.

First, the anomalous contribution to the specific heat could be due to non-Fermi liquid electronic degrees of freedom. Well known examples of groups of materials in which the possibility that part of the electronic system can be described by fermionic itinerant quasiparticles and others not is considered theoretically, are the

5.1 Caloric Studies of Magnetic Phase Transitions in Sr$_3$Ru$_2$O$_7$

normal state of the cuprate high-T_C superconductors [13] or Kondo systems in the regime in which they contain coexisting localized and itinerant electrons [14]. However, I am not currently aware of microscopic theories for Sr$_3$Ru$_2$O$_7$ predicting additional degrees that result in the particular form of the observed specific heat.

The second scenario is that the T^2 term in the specific heat has its origin in finite temperature corrections to Fermi liquid theory. One possible theory discussed by Chubukov et al. [15] describes non-analytic quantum field corrections to Fermi liquid theory. Here the authors predict that the specific heat c divided by temperature T of a two-dimensional Fermi liquid is given by

$$\frac{c}{T} = \gamma - B\left(\frac{m^*}{k_F}\right)^2 T\left(f_C^2(\pi) + 3f_S^2(\pi)\right), \tag{5.9}$$

where γ is the usual Sommerfeld coefficient and $f_C(\pi)/f_S(\pi)$ are the backward scattering amplitudes in the charge (C) and spin (S) channel. Discussing the details of the theory is beyond the scope of this thesis. The first important aspect relevant to the discussion here is that it predicts a correction to the specific heat of a two-dimensional Fermi liquid that is quadratic in temperature. It is furthermore quadratic in the effective mass. It was also shown in [15] that $f_S(\pi)$ depends crucially on the Wilson ratio, with large Wilson ratios resulting in large backward scattering amplitudes in the spin channel. Sr$_3$Ru$_2$O$_7$ is a material that first of all has a quasi-two-dimensional Fermi surface. Secondly its Fermi surfaces have relatively large effective masses m^* of the order of 10. More importantly the material has a zero field Wilson ratio R_W of the order of 10 [16]. It is therefore at least a possibility that the anomalous specific heat contribution represents a finite temperature correction to Fermi liquid theory. However, the data discussed in the next section regarding the 'normal' states surrounding the novel quantum phase show that the quadratic correction to the specific heat is only measureable in the novel quantum phase and is therefore, if existing, significantly smaller in the surrounding 'normal' states. In order to reconcile this with the theoretical model by Chubukov et al. one has to assume that the backward scattering amplitudes in the charge or spin channel, $f_C(\pi)$ or $f_S(\pi)$, are significantly enhanced in the anomalous phase compared to outside the phase, since the difference in effective mass between the phases is only of order one. This would indicate a fundamental difference in the residual interactions between the Landau Fermi liquid quasiparticles in the novel quantum phase compared to the low and high field states of Sr$_3$Ru$_2$O$_7$.

Finally it should be mentioned that several theoretical proposals for the nature of the novel phase based on microscopic properties of Sr$_3$Ru$_2$O$_7$ have been put forward. Examples are spin-spirals [2], Pomeranchuk-type instabilities of the Fermi surface [17, 3] and the formation of electronic 'nematic-like' phases [18, 19]. Though these theories in general predict the overall topology of the phase diagram and are in principle consistent with anisotropic transport by showing a change in symmetry of the Fermi surface, none of them predict the above discussed corrections to the Fermi liquid specific heat nor the curvature of the first order phase transition lines.[9]

5.2 The Low and High Field States of $Sr_3Ru_2O_7$

5.2.1 The Low Field Fermi Liquid State

In this section the results of caloric studies of the low field state will be presented. In particular the question of what happens to the entropy and specific heat upon approaching the critical region will be discussed. The second part of the section then concentrates on observed magnetocaloric oscillations in order to further investigate the Fermi liquid contribution to the entropy of the system.

5.2.1.1 Divergence in Entropy and Specific Heat

As discussed in the previous section it is possible to reconstruct the entropy change as a function of magnetic field from the measurements performed in this work. In Fig. 5.10 the entropy change ΔS relative to 5 T divided by temperature T is shown as a function of magnetic field. The traces are offset for clarity where the offset scales with the temperature difference between the traces.

The most striking feature of the entropy in the low field state is the rapid increase towards the critical region. In previous work, both experimentally [8] and

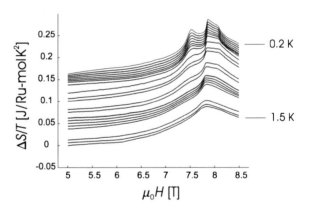

Fig. 5.10 Here the entropy change ΔS relative to 5 T divided by sample temperature T as a function of magnetic field H is shown for several temperatures between 0.2 K and 1.5 K. The entropy scale is that of the 1.5 K curve. The other data have the same scale but are offset in proportion to the difference in sample temperature

---

[9] Though the experimentally observed 'nematic-like' resistivity properties of the anomalous novel phase are not a direct subject of this thesis it should be noted that no explicit calculation of the transport properties has been carried out to my knowledge.

5.2 The Low and High Field States of $Sr_3Ru_2O_7$

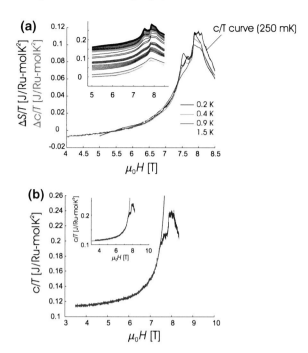

Fig. 5.11 **a** The change in entropy ΔS with respect to 5 T divided by temperature as a function of magnetic field. The sample temperature is 0.2 K (*black*), 0.4 K (*green*), 0.9 K (*red*) and 1.5 K (*yellow*). In the *inset* the previously shown graph of Fig. 5.10 is shown with the position of the *curves* highlighted in the same colours. Furthermore, overlaid in *blue* is the change relative to 5 T of the specific heat Δc divided by temperature T as a function of magnetic field. Here the sample temperature is 250 mK. N.B. both quantities, entropy and specific heat, have the same units and scale. In **b** the specific heat trace of **a** is shown together with a power law fit with an exponent of -1 (*red curve*). In the *inset* the same specific heat trace is shown. However, this time the *red curve* represents a simulation of the specific heat based on a Fermi liquid single particle density of states peak having a power law singularity with an exponent of $-1/2$. For further details see text

theoretically [20] it has been argued that upon approaching the critical region in magnetic field from the low field Fermi liquid state the quasiparticle effective masses should increase for at least part of the quasiparticle excitation spectrum. Such an increase in effective mass would cause an increase in entropy and low temperature specific heat upon approaching the critical region from the low field side as observed.

In Fig. 5.11 the field and temperature dependence is analysed in detail. Part (a) shows four traces of the change in entropy S divided by temperature T as a function of magnetic field for the temperatures 0.2 K (black), 0.4 K (green), 0.9 K (red) and 1.5 K (yellow). The inset shows these curves highlighted in the graph shown in the previous figure. Furthermore in blue is shown a measurement of the AC specific heat c divided by temperature T as a function of field for a sample temperature of 250 mK. The curve is offset such that its zero is at 5 T and therefore represents the change in c/T relative to this field.

The entropy curves are in agreement (within experimental error) between 5 and 7.2 T. It is therefore reasonable to assume that over the temperature and field range represented by these curves the entropy is consistent with a Fermi liquid state that has an increasing quasiparticle mass with increasing field. This is furthermore confirmed by the agreement of $(S(H) - S(5T))/T$ with $(c(H) - c(5T))/T$ since for a Fermi liquid at low temperature the identity $S/T = c/T = const$ holds.

Since the relative accuracy of the AC specific heat measurement is better than that of the magnetocaloric sweeps it is this curve that is used for the further analysis of the divergence in entropy.

It was found that the behaviour is best described by a power law divergence of the form

$$c/T = c_0/T + b \left[\frac{H - H_C}{H_C} \right]^d. \quad (5.10)$$

This fit (red curve) is shown together with the specific heat trace (blue curve) in Fig. 5.11b. The fit has been performed over the maximum field range from 3.5 to 7.4 T and parameters are the non-singular contribution to the specific heat $c_0/T = 106.5(5)$ mJ/Ru-mol K^2, the critical field $H_C = 7.90(5)$ T, the exponent $d = -0.98(10)$ and the prefactor $b = 5.3(1)$ mJ/Ru-mol K^2.

This result indicates that the singular part of the entropy is diverging with a power law consistent with the power -1. However, one has to be careful in the interpretation of this exponent. One scenario for example would be that the peak in entropy is caused by a peak in the single particle density of states of the band structure of Sr$_3$Ru$_2$O$_7$. In this case applying the magnetic field would split the Fermi surface into a spin-up and spin-down contribution with the spin-down surface for example being moved through the density of states peak. However, due to number conservation the distance of the chemical potential $\mu_\downarrow$ surface from the peak position e_C in the density in states, i.e. $(\mu_\downarrow - e_C)/e_C$ is not proportional to the distance in magnetic field from the critical field, i.e. $(H - H_C)/H_C$ as has been discussed in Sect. 2.1.2. Important for the discussion here is that for the particular value of -1 of the power law divergence in the underlying density of states singularity would have to diverge with a power of $-1/2$. In other words the singular part of the density of states would have to be proportional to

$$\left(\frac{\mu_\downarrow - e_C}{e_C} \right)^{-1/2}. \quad (5.11)$$

This is shown in the inset of Fig. 5.11b where in blue the measured specific heat is shown together with the calculated specific heat (in red) for an example density of states that has a singular part with an exponent of $-1/2$.

In summary, the measurements show that the low temperature specific heat is consistent with that of a Fermi liquid whose density of state is increasing as a function of magnetic field and has a singular contribution at a critical field of 7.9 T. However, it should be emphasised that it is not possible to draw any firm

5.2 The Low and High Field States of $Sr_3Ru_2O_7$

conclusion upon the cause of the increase in entropy towards the critical region. The observed behaviour can be interpreted within several scenarios, such as quantum criticality and an unusual density of states peak. A more detailed discussion will be given at the end of this section.

5.2.1.2 Quantum Oscillations in the Magnetocaloric Signal

One feature of a Fermi liquid is the possible observation of quantum oscillation phenomena at low temperature and high magnetic fields. As discussed with the example of Sr_2RuO_4 in Sect. 4.4.4, it is possible to observe these quantum oscillations in the magnetocaloric sweeps and gain information on the Fermi surface such as size and effective mass.

Figure 5.12a shows an example of a magnetocaloric sweep at $T_{bath} = 150$ mK. Here the blue curve is a measurement of the sample temperature while increasing the magnetic field at 0.04 T/min and the red curve while decreasing the magnetic field at −0.04 T/min.[10] The oscillatory component is out of phase by 180° as expected for quantum oscillations. An in-depth study of quantum oscillatory phenomena in $Sr_3Ru_2O_7$ has been done using the de Haas–van Alphen effect by J.-F. Mercure and co-workers [12]. In Fig. 5.12b a comparison is shown of the frequency spectrum observed in this work[11] (black curve) with a spectrum

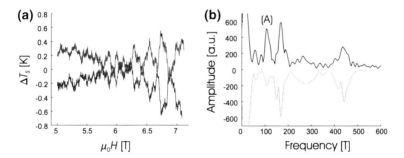

Fig. 5.12 **a** is showing the change of sample temperature ΔT_S during a sweep of the magnetic field at 0.04 T/min (*blue*) and −0.04 T/min (*red*) as a function of magnetic field H. The average temperature is 160 mK. **b** shows in *black* the Fourier transform in $1/H$ of the *blue trace* in **a**. In *green* is given the Fourier transform of a de Haas–van Alphen measurement over a comparable magnetic field region at base temperature. The samples used in both experiments were of similar quality. The label *A* indicates the new frequency found at 110 T

---

[10] All measurements on magnetocaloric oscillations presented in this work were performed on sample C698K with a mass of 23 mg. This is the same sample as used for the angular study reported later on. Quantum oscillations were observed in all samples used during the measurements.

[11] The temperature of the sample for this spectrum is 150 mK and the field range over which the Fourier transform was done is 3.6 T–7.1 T.

obtained at base temperature in the de Haas–van Alphen experiments (green curve) over a similar field range (the sign of the latter spectrum is changed for clarity). The two quantum oscillation experiments are in good agreement in their frequency spectrum. At higher frequencies the magnetocaloric oscillations are thermally damped and the de Haas–van Alphen experiment has a far better sensitivity. However, the magnetocaloric effect is in particular sensitive to oscillations at low frequencies due to a low background contribution and the above mentioned damping of higher amplitude high frequency oscillations. In fact the frequency at 110 T [labelled (A) in the graph] was first clearly observed above the noise level in the magnetocaloric measurements. The frequency was absent in a test experiment of a magnetocaloric sweep without a sample, confirming it to be part of the $Sr_3Ru_2O_7$ quantum oscillation spectrum. After an optimisation of the experimental parameters it was subsequently confirmed by de Haas–van Alphen measurements.

In order to establish the contribution to the specific heat of this part of the Fermi surface a temperature study of the oscillations was performed. This was limited by the highest temperature at which oscillations were still sufficiently observable (approximately 200 mK) and the lowest temperature achievable with the current setup (90 mK). Figure 5.13 shows the amplitude of the oscillations as a function of temperature for measurements of both increasing and decreasing magnetic field sweeps. The amplitude was obtained from a Gaussian fit to the frequency spectrum between 80 and 130 T. The red curve represents a fit of the temperature derivative of the Lifshitz-Kosevich formula for the temperature dependence of the oscillations (see Sects. 3.2.2, 4.4.4). From this fit an effective mass of $8(1)m_e$ is deduced, with m_e being the free electron mass. The result was subsequently confirmed by a detailed de Haas–van Alphen study [12]. The size of the pocket together with the effective mass makes plausible an identification of this part of the Fermi surface with the γ_2 pocket as reported by Tamai et al. in ARPES measurements (see Sect. 2.2.3). According to their measurement, this pocket occurs four times in the first Brillouin zone and is furthermore most probably bilayer split. Any contribution based on the measured effective mass has therefore to be multiplied by a factor of eight to give the total specific heat contribution of this part of the Fermi surface.

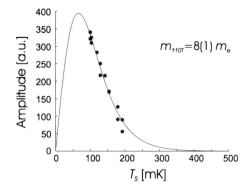

Fig. 5.13 The data in *blue* gives the amplitude of the 110 T peak in the Fourier transform of the data in **a** as a function of sample temperature T_S. In *red* is shown a fit of the temperature derivative of the Lifshitz–Kosevich function to the data. This fit gives a mass $m_{110\,T}$ for the newly observed frequency of $8(1)m_e$, with m_e being the free electron mass

5.2 The Low and High Field States of $Sr_3Ru_2O_7$

For a quasi-two-dimensional Fermi surface, the specific heat contribution is directly proportional to the effective mass of an orbit. Here the newly observed frequency is in total contributing 47(6) mJ/Ru-mol K^2. The other five orbits as measured by J.-F. Mercure contribute a further 56(1) mJ/Ru-mol K^2, giving a total of 103(7) mJ/Ru-mol K^2. This is very close to the measured average of 120 mJ/Ru-mol K^2 over the field range of 4–7 T as shown in Fig. 5.11b. It therefore appears that the low frequency observed in this experiment is contributing approximately 45% of the total specific heat.

In this section it was first of all shown that magnetocaloric oscillations are observable in the samples used for this project. Furthermore, this way of measuring quantum oscillations is particularly sensitive to low frequencies and revealed for the first time a clear signature of a small Fermi surface contributing a significant part to the overall specific heat.

5.2.2 The High Field Fermi Liquid State

In this section results on the entropic properties of the high field state above the critical field will be presented.

5.2.2.1 Magnetocaloric Measurements

The analysis of the data for the high field phase of $Sr_3Ru_2O_7$ above the critical phase was done analogously to the data presented in the previous section. Figure 5.14 shows the isothermal change of entropy as a function of magnetic field with respect to 5 T. The curves have been offset as previously from each other for clarity, the offset scaling with the temperature difference.

The most significant feature is, as for the low field side, the strong increase in entropy upon approaching the critical field region. Also clearly identifiable is the entropic signature of a previously reported metamagnetic crossover at approximately 12.5 T [21]. Finally, clearly observable are low temperature quantum oscillations. The combination of these features makes a detailed quantitative analysis of the divergence in entropy as in the previous section impossible. However, in the field region from 8.2 to 10 T the data are consistent with the increase in entropy being such that, for a given magnetic field, S/T can be assumed to be a constant of temperature over the phase space explored by the experiment. Therefore, the conclusion that the observations are consistent with the high field state being a Fermi liquid in this region of magnetic field follows equivalently. Above 10 T deviations assumed to be associated with feature (A) become important.

Finally, I would like to draw attention to the oscillatory component at low temperatures. These again are quantum oscillations being directly observed in the entropy of the system. The observed oscillations are discussed in more detail in the following section.

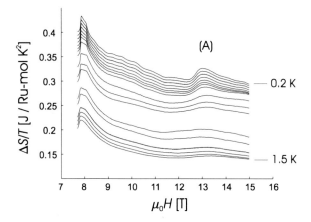

Fig. 5.14 In this figure, analogous to Fig. 5.10, the entropy change ΔS relative to 5 T divided by sample temperature T is shown as a function of magnetic field H. The curves correspond to several temperatures between 0.2 K and 1.5 K. The entropy scale is the same for all curves, which are offset in proportion to the difference in sample temperature (there is no offset for the 1.5 K curve). The label A indicates the position of a further metamagnetic crossover which has been reported previously [21]

5.2.2.2 Quantum Oscillations in the Magnetocaloric Signal

Figure 5.15a shows a trace of the sample temperature during a magnetic field sweep of 0.04 T/min (blue) and −0.04 T/min (red) with a second order polynomial background subtracted. The temperature of the thermal bath was 160 mK.

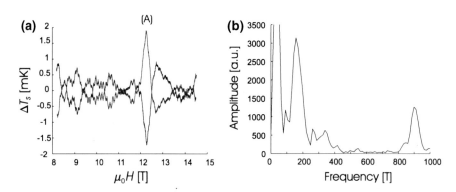

Fig. 5.15 **a** The change in temperature relative to the average temperature during a sweep of the magnetic field at a rate of 0.04 T/min (*blue*) and −0.04 T/min (*red*) on the high field side of the critical field region. The feature labelled with (A) is not part of the oscillatory signal but the magnetocaloric signature of a thermodynamic crossover. The sample average temperature is 160 mK. **b** The frequency spectrum given by the Fourier transform in inverse magnetic field $1/H$ of the *blue trace* shown in **a**

The strong feature at 12.5 T is not part of the oscillatory signal but the previously mentioned signature of a metamagnetic crossover. Figure 5.15b shows the Fourier transform of the signal for the frequency region where the experiment is most sensitive. There is significant spectral weight between 100 and 200 T. Unfortunately it is not possible to resolve if this is a combination of several frequencies or if, for example, the 110 T frequency corresponding to the γ_2 surface has vanished. A temperature analysis of the oscillations showed the measured effective masses to be consistent with the de Haas–van Alphen measurements by Mercure et al. [12].

5.2.3 Discussion

In this section results of magnetocaloric and specific heat measurements in the low and high field phases of $Sr_3Ru_2O_7$ have been presented. Figure 5.16 shows the overal results for the entropy S divided by temperature T (Fig. 5.16) and the specific heat c divided by T (Fig. 5.16) as a function of magnetic field H and temperature T, including the novel quantum phase. Both quantities, the entropy

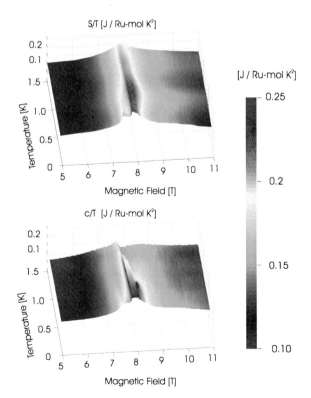

Fig. 5.16 Smoothed surfaces of the entropy S and specific heat c of $Sr_3Ru_2O_7$ divided by temperature T as a function of magnetic field and temperature. The lowest temperature shown in the data is 200 mK. The colour scale is identical for both quantities. The entropy surface is based on the assumption that $Sr_3Ru_2O_7$ has a Fermi liquid like specific heat at 5 T, justified by c/T being constant at that field as shown

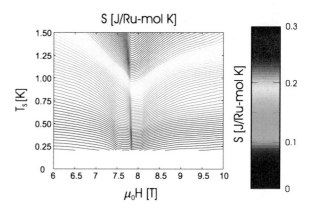

Fig. 5.17 The isoentropes of $Sr_3Ru_2O_7$ in the $H - T$ plane. The corresponding colour scheme is given on the right hand side

and the specific heat, show a pronounced peak structure centred on the critical field of 7.9 T.

By plotting both the entropy and the specific heat divided by temperature one emphasises in particular the Fermi liquid like character of the low and high field phases, since both quantities are consistent with being constant as a function of temperature in these regimes.[12]

However, it is equally informative to plot the entropy itself. This is done in Fig. 5.17 in form of equidistant isoentropes of $Sr_3Ru_2O_7$ in the $H - T$ diagram.

In principle, the isoentropic plot contains exactly the same information as the plot in Fig. 5.16. The physical meaning of these curves is that the temperature of the sample would evolve along them in a perfectly adiabatic magnetocaloric measurement. Upon approaching the critical region from the low or high field side the entropy is 'dragged' towards lower temperature, supposedly resulting in a singularity at the critical field. That this does not violate the third law of thermodynamics can be inferred from the fact that at the same time isoentropes at lower temperature have a smaller curvature than isoentropes at higher temperature, permitting $S(T = 0) = 0$ to hold for the low and high field states of $Sr_3Ru_2O_7$ at all fields.

The three thermodynamic features, i.e. the two first order transitions (1) and (2) as well as the thermodynamic crossover (3) are clearly identifiable. Both the isoentropes between the thermodynamic crossover (3) and the phase transition (1) as well as the isoentropes inside the anomalous phase have a much weaker curvature than in the the low and high field states of $Sr_3Ru_2O_7$.

From the qualitative features one is therefore tempted to assume the physics of $Sr_3Ru_2O_7$ to be driven by a possible singularity in the entropy at zero temperature with the material undergoing a series of transitions as a function of field in order to

---

[12] The deviations as a function of temperature in the high field state are due to systematic errors when integrating up the magnetocaloric traces as a function of magnetic field. Since this integration is done with reference to 5 T a small random error in any of the quantities can accumulate over the integration range.

Fig. 5.18 Reproduction of Fig. 2.19c. This shows the Fermi surface of $Sr_3Ru_2O_7$ in the k_x–k_y plane as measured by ARPES [24] with the first Brillouin zone given by the blue diamond outline. M, X and Γ are points of high symmetry in the Brillouin zone. Also labelled are the Fermi surfaces as explained in the text

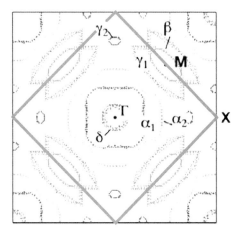

avoid this singularity. However, the wealth of experimental information, in particular of the zero field phase of $Sr_3Ru_2O_7$ and its evolution as a function of magnetic field, make a far more quantitative discussion possible. In the following I will therefore first analyse the extent to which the single particle density of states of the zero field Fermi liquid could explain both qualitatively and quantitatively the observed peak in entropy. In a second part I will then discuss how other phenomena which are expected to play a significant role in the material, such as for example quantum criticality, can change the properties of $Sr_3Ru_2O_7$ as a function of magnetic field.

First, I will summarise the current experimental evidence for the electronic structure of the low field state. Several authors have conclusively shown the existence of quantum oscillations in both the low field and high field state of $Sr_3Ru_2O_7$ [22, 23, 12]. The inferred existence of a Fermi surface and quasiparticle excitations has furthermore been confirmed at zero field by angular resolved photoemission spectroscopy (ARPES) [24]. In Fig. 5.18 the Fermi surface in the k_x/k_y plane as reconstructed from ARPES measurements is shown (this is the same figure as 2.19c in Sect. 2.2.3). The angular dependence of the frequencies in quantum oscillation measurements show that these Fermi surface sheets are quasi-two-dimensional in character with their primary axis being parallel to k_z. The main characteristics of the zero field Fermi surface as deduced from these measurements were presented in Table 2.1 in Sect. 2.2.3 and reproduced here in Table 5.1. This summary also includes the γ_2 pocket whose existence at the Fermi energy was first demonstrated conclusively by the study presented in this thesis.

The ARPES measurements together with local density approximation (LDA) calculations allowed for an identification of the multiplicity of each Fermi surface part in the Brillouin zone [24]. This information is given in the last two rows of Table 5.1. Here 'Symmetry' refers to the multiplicity due to the symmetry of the Brillouin zone. The γ_2 pocket for example occurs four times. Furthermore 'Bilayer Split' refers to the bilayer split of some Fermi surfaces that has to be taken into account when calculating the physical properties of the material, but, if small,

Table 5.1 The properties of the $Sr_3Ru_2O_7$ Fermi surfaces as measured by quantum oscillation and ARPES experiments [12, 24]

	α_1	α_2	β	γ_1	γ_2	δ
dHvA						
F (kT)	1.78	4.13	0.15	0.91	0.11	0.43
Area (% BZ)	13.0	30.1	1.09	6.64	0.31	3.14
m^* (m_e)	6.9 ± 0.1	10.1 ± 0.1	5.6 ± 0.3	7.7 ± 0.3	8 ± 1	8.4 ± 0.7
ARPES						
Area (% BZ)	14.1	31.5	2.6	8.0	< 1	2.1
m^* (m_e)	8.6 ± 3	18 ± 8	4.3 ± 2	9.6 ± 3	10 ± 4	8.6 ± 3
Symmetry	1	1	2	2	4	1
Bilayer split	1	1	1	2	4	2

F is the observed frequency in $1/H$ of quantum oscillations, BZ stands for Brillouin zone and m^* is the effective mass in units of electron mass m_e. This table contains the same information as Table 2.1 in Sect. 2.2.3 with the addition of the effective mass of the γ_2 pocket first conclusively measured in the work presented in this thesis

would be experimentally unresolvable by quantum oscillation measurements. Together with the γ_2 pocket the specific heat calculated from the effective masses of all Fermi surfaces assuming the above multiplicities is 103(7) mJ/Ru-mol K^2 and consistent with the experimentally measured zero field specific heat of 110 mJ/Ru-mol K^2 [5].

A very important result from the quantum oscillation measurements is the measured field dependence of the effective mass m^* of the identified frequencies. Since m^* for each quasi 2D Fermi surface is directly proportional to its contribution to the overall specific heat one would expect at least part of the Fermi surface to show a significant peak in its effective mass in the vicinity of the critical field of 7.9 T if this peak is caused by quasiparticle excitations. Mercure et al. showed that one can exclude this peak in effective mass for all of the identified frequencies except that corresponding to the γ_2 pocket, for which the analysis was not possible. In other words, these Fermi surface parts cannot explain the observed peak in the specific heat. Therefore, any explanation of the metamagnetism in $Sr_3Ru_2O_7$ based primarily on the zero field density of states has to assume that γ_2 is the Fermi surface pocket at the centre of the unusual physical effects observed.

Importantly, ARPES measurements by Tamai et al. [24] showed that the electronic band associated with the γ_2 Fermi surface pocket has a density of states peak associated with a saddle point energetically close to the Fermi energy. The authors argue that this can in principle be the cause for the observed metamagnetism by spin-splitting the Fermi surface and shifting the chemical potential for one of the spin species through the observed density of states peak.

The most important question arising from the data is therefore if the zero field band structure $Sr_3Ru_2O_7$ can both qualitatively and quantitatively account for the observed peaks in entropy and specific heat presented in this thesis. In a first scenario I will assume the hypothesis that the observed singular contribution in the entropy S divided by temperature T at the critical field of 7.9 T is indeed due to the band structure at zero field. I will discuss the available experimental data under the constraint of this assumption and point out possible inconsistencies. It has to be mentioned here that this part of the discussion is primarily based on the assumption that indeed all relevant Fermi surface parts of $Sr_3Ru_2O_7$ have been identified and their multiplicities correctly assigned. In the second part of the following discussion possible other contributions to the peak in entropy will be presented. In particular the possibilities of either a change in the single particle density of states as a function of magnetic field (as for example predicted by theories for quantum criticality in $Sr_3Ru_2O_7$) or of non-Fermi liquid degrees of freedom will be addressed.

5.2.3.1 Single Particle Density of States Peak

ARPES measurements such as those performed by Tamai et al. [24] are among the few experimental methods able to probe the density of states of a Fermi liquid not only at the Fermi surface but also as a function of energy below the Fermi energy. In Fig. 5.19 the measured density of states for the energy bands resulting in the γ_1 and γ_2 Fermi surfaces as well as the sum total are shown as a function of energy relative to the Fermi energy.[13]

The specific heat at zero field of 110 mJ/Ru-mol K^2 corresponds to 47 states/Ru eV [24]. Of these, 26 states/Ru eV are accounted for by the five Fermi surfaces α_1, α_2, β, γ_1 and δ. These five bands cross the Fermi energy ϵ_F such that the density of states does not vary significantly within several meV of ϵ_F, consistent with the observed weak changes in effective quasiparticle masses in de Haas–van Alphen measurements [12]. γ_2 on the other hand not only is close to a van Hove-singularity

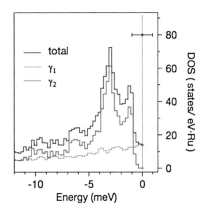

Fig. 5.19 Density of states as a function of energy of the bands associated with the γ_1 and γ_2 Fermi surface as well as the sum total. The figure differs from that in the original publication by Tamai et al. due to a corrected error that overestimated the density of states of the γ_1 associated band by a factor 2

---

[13] This is not the graph originally published by the authors. The data shown here is corrected for an error in the density of states for the γ_1 band which was overestimated by a factor of 2.

but also to the top of the band. Therefore, the density of states of the band associated with the γ_2 pocket varies rapidly within ±5 meV of the Fermi energy. A combination of uncertainty in the exact position of the Fermi energy of ±1 meV and a possible relative shift of the bands due to surface effects[14] makes it impossible to perform an accurate calculation of the exact contribution to the specific heat on the basis of the ARPES measurements alone. However, from the magnetothermal oscillations presented in this thesis it is possible to extract the expected specific heat contribution to be of the order of 47(6) mJ/Ru-mol K^2 or 22 states/Ru eV. Therefore, the Fermi energy is expected to lie at least 1–2 meV below the one shown in Fig. 5.19. Here it is important to note that it is currently not possible to study a potential effective mass divergence of the γ_2 Fermi surface as a function of magnetic field. This is due to the fact that in order to sufficiently resolve this low frequency one has to Fourier transform the oscillation signal over the whole range of available magnetic field from 4 T to 7 T. It remains to be seen if these difficulties can be overcome in future high resolution measurements in combination with advanced data analysis.

The maximum height of the density of states of the band associated with the γ_2 pocket is of the order of 60 states/Ru eV including spin degeneracy, 40 states/Ru eV above the zero field value. This implies that the maximum number of states per Ru per eV that can be gained per spin species is 20 states/Ru eV. In the most simple scenario, applying a magnetic field lifts the degeneracy in spin resulting in the spin-up band being lowered in energy while the spin-down band is raised in energy. This would lead to the spin-up band to being such that the Fermi energy can pass through the density of states peak, gaining at most of the order of 20 states/Ru eV in density of states. The spin-down band on the other hand has a density of states that in the most simple scenario will be constant to first order or drop significantly if the top of the band is reached. Therefore, the zero field density of states as measured by ARPES can at most account for an increase of 20 states/Ru eV in the density of states or 47 mJ/Ru-mol K^2 in the Sommerfeld coefficient as a function of magnetic field. The specific heat measurements presented in this thesis on the other hand show an increase in the Sommerfeld coefficient of the order of 100 mJ/Ru-mol K^2 up to a magnetic field of 7 T. Therefore, a calculation of the increase of the specific heat as a function of field based on the zero field density of states as measured by ARPES underestimates the experimentally measured increase in the specific heat by a factor of 2. This can be assumed to be a large enough discrepancy to conclude that the simple model presented so far, based on the zero field density of states of Sr$_3$Ru$_2$O$_7$ as measured by ARPES, is inconsistent with the experimental observations. There are several possibilities however how to reconcile the zero field ARPES and field dependent specific heat measurements.

---

[14] ARPES is not a bulk but surface probe. It is always possible that a small surface reconstruction results in small changes of the band structure at the surface relative to the bulk. This however, is assumed in first order to only shift the bands relative to each other but not actually to affect the amplitude of the single particle density of states.

5.2 The Low and High Field States of $Sr_3Ru_2O_7$

First, it could be that the dominant effect of applying the magnetic field is not a splitting of the spin-up and spin-down Fermi surface but a shift of both spin-up and spin-down bands through the van Hove singularity. However, such a shift is not a direct result of a simple single particle band structure effect and would require a rigorous theoretical justification.

A further possibility is a second density of states peak just above the Fermi energy, not being measureable by ARPES. In this case the spin-down bands could also contribute to the specific heat enhancement. Indeed, scanning tunneling microscopy experiments by Iwaya et al. [25] seemed to indicate the existence of two peaks above and below the Fermi energy in the local density of states as measured by the current-voltage characteristic of the tunneling current. However, the density of states extracted in their measurements at the Fermi energy as a function of magnetic field did not show any discernable peak around 8 T but simply a monotonic increase over the measured field region. Therefore, their results are fundamentally inconsistent with the interpretation of an electronic single particle density of states peak causing the observed peak in entropy and specific heat as a function of magnetic field. It is consequently not directly obvious how the features observed by Iwaya et al. can be related to the band structure.

One further possible experiment to probe the band structure above the Fermi energy is by electron doping. In the strontium ruthenate family this can be achieved by replacing Sr atoms in the 2+ oxidation state chemically with La atoms that are in the 3+ oxidation state. This was successfully achieved by Kikugawa et al. in the case of Sr_2RuO_4 [26]. In this study a series of $Sr_{2-y}La_yRuO_4$ samples were prepared and the authors succesfully showed that the effect of La doping is filling the original band structure of Sr_2RuO_4 with electrons, in a so-called 'rigid

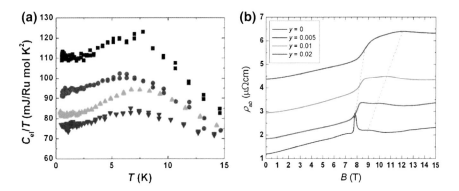

Fig. 5.20 Figures reproduced from Farrell et al. [27]. In the *left panel* the electronic specific heat C_{el} divided by temperature T as a function of T is given for different doping levels of La in $Sr_3Ru_2O_7$ as detailed in the *inset* in **b**. In the *right panel* the in-plane magnetoresistance ρ_{ab} as a function of magnetic field is given for the same doping levels. The field is applied along the crystallographic c-axis. The *broken lines* represent the trend of the metamagnetic features with doping

band shift' of the Fermi level. A similar study for $Sr_3Ru_2O_7$ has been performed by Farrell et al. [27].

In Fig. 5.20 the most important results are presented. First, as is shown in (a), with small amounts of doping the zero field Sommerfeld coefficient rapidly falls from 110 mJ/Ru-mol K^2 with no doping to 78 mJ/Ru-mol K^2 at $y = 0.02$. Furthermore, one observes that with increased doping the metamagnetic features shift to higher fields as is shown in Fig. 5.20b. Here the in plane resistivity ρ_{ab} is shown as a function of magnetic field for several doping levels as indicated in the inset. The broken lines indicate the trend for the features associated with the metamagnetic transitions to shift to higher fields with higher doping. The broadening in these features is ascribed to impurity scattering. In the scenario discussed here it is assumed that La doping is simply adding electrons to the band structure and thereby shifting the Fermi energy through the single particle density of states, as it does in the case of Sr_2RuO_4. Therefore, from the first observation of the sharp drop in the Sommerfeld coefficient one would conclude that the Fermi energy ϵ_F is very close to sharp drop in the density of states above ϵ_F. This is qualitatively consistent with the observation in ARPES that the Fermi energy is close to the top of the band associated with the γ_2 pocket. From the second result, the shift of the metamagnetic features to higher magnetic fields with higher doping, one would conclude that the dominant feature in the single particle density of states causing the metamagnetic transitions is situated below the Fermi energy, again consistent with the ARPES data as discussed before. This does not exclude a further density of states peak above the Fermi energy since it only indicates that the dominant feature is below the Fermi energy.

However, the data also include features that contradict a simple interpretation of the doping study. The most significant is a pronounced peak in the zero field specific heat at 8 K as shown in Fig. 5.20a. In a simple band structure model this peak is associated with the density of states peak in the band structure as follows. At finite temperature T the Fermi distribution of electrons does not have a sharp discontinuity at the Fermi energy but changes smoothly over a range in energy of the order $k_B T$. Therefore, if there is a density of states peak at an energy $\Delta \epsilon_F$ away from the Fermi energy one would expect the Fermi–Dirac distribution to sample that density of states peak at temperatures of the order of that energy scale. This leads to a peak in the density of states. If indeed the Fermi energy is moved significantly through the band structure, as seen by the shift in metamagnetic field, than the zero field density of states peak would also be expected to shift consistently. In [27] it is argued that in a simple band structure picture this shift would at least be of the order of 10 K to higher temperatures for a La doping level of $y = 0.06$. This, however, is not consistent with the data shown in Fig. 5.20a. It follows that the simultaneously observed shift of the metamagnetic critical field and the unchanged position in temperature of the zero field specific heat peak as a function of La doping are seemingly contradictory features if interpreted within the assumption of a simple shift of the Fermi energy through a rigid single particle density of states. This issue will have to be addressed by further detailed studies of these samples by, for example, ARPES.

5.2 The Low and High Field States of Sr$_3$Ru$_2$O$_7$

Finally, the observed functional form of the peak in the low temperature specific heat c of Sr$_3$Ru$_2$O$_7$ has to be discussed. As has been shown in Fig. 5.11b the increase in specific heat c in the low field state is consistent with a singular contribution to c, which is proportional to

$$(H - H_C)^{-1}. \tag{5.12}$$

As discussed previously this implies that if the peak in specific heat is dominated by a peak in the density of states $g(\epsilon)$ of the band structure at a critical energy ϵ_C, then the singular contribution to $g(\epsilon)$ is proportional to

$$(\epsilon - \epsilon_C)^{-1/2}. \tag{5.13}$$

In a three dimensional band structure no true divergence of the density of states is possible. However, in two dimensions logarithmic divergences at saddle points are possible. Therefore, if the band structure of a solid state material is quasi-two-dimensional, it is possible that the density of states in the vicinity of a saddle point peaks sharply, with any divergence ultimately being cut off due to the true three dimensionality of the material. If one assumes that this is the case in Sr$_3$Ru$_2$O$_7$ then one would expect the observed divergence to be of logarithmic form as shown in Fig. 5.21a, but not the particular power law given in 5.13. Attempts to construct a density of states with a logarithmically divergent contribution whose magnetic field dependent specific heat agrees with the experimental results on the same level as the density of states with a power law divergence have not been successful.

In order to reconcile these two points, one has to assume that the peak in the density of states looks like a power law over the energy range probed here. In this

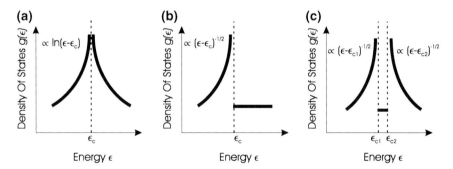

Fig. 5.21 In this figure possible scenarios for divergences in the density of states $g(\epsilon)$ as a function of energy arising generically from band structure features are given. **a** In two dimensions $g(\epsilon)$ can have a singular part proportional to a logarithmic divergence at a critical energy ϵ_C. The divergence is symmetric in field. **b** For one dimensional bands the density of states can diverge at the top of a band as shown. This divergence is proportional to a power law with exponent $-1/2$ but asymmetric, since there are not states above the band edge. The same analogous divergence can appear at the bottom of a band. If the band structure is such that the top of a one dimensional band is at a critical energy ϵ_{C1} just below the bottom of a one dimensional band at a slightly larger energy ϵ_{C2}, then a symmetric situation can arise in a band structure as shown in **c**. Here the density of states diverges in the form of a power law with the exponent $-1/2$ on both sides

case one has to postulate that this form is not due to a generic effect but to particular circumstances in $Sr_3Ru_2O_7$. It is possible in quasi-one-dimensional band structures at the top of a band to have a singular contribution to the density of states of the form of 5.13. However, as shown in Fig. 5.21b this kind of density of states shows the power law divergence only on one side of the singularity, whereas the other side is constant in energy. Therefore, in order to explain the symmetric peak in entropy as measured in this experiment one would have to postulate two quasi-one-dimensional bands in $Sr_3Ru_2O_7$ that are such that energetically the top of one of them is very close to the bottom of the other (Fig. 5.21c). This again would require the postulate of a very specific form of the band structure in order to explain the functional form of the peak in the density of states.

In summary the observed symmetric power law divergence in entropy as a function of magnetic field is not a feature that arises generically from a van Hove singularity in a quasi-two or one-dimensional band structure. One possibility is to postulate a particular non-generic band structure of $Sr_3Ru_2O_7$ that results in the observed specific heat as a function of magnetic field. The other possibility is that the functional form of the observed peak in the entropy cannot be explained only in terms of a band structure feature crossing the Fermi energy.

An alternative possibility discussed in the following is that the zero field density of states as measured by ARPES is significantly changed as a function of field due to field dependent quasiparticle interactions.

5.2.3.2 Alternative Scenarios

An alternative scenario to a rigid shift of the spin-dependent density of states as a function of magnetic field is that the zero field density of states as measured by ARPES is significantly changed due to magnetic field dependent quasiparticle interactions.

One possibility of enhancing the density of states of an electronic band structure is a renormalisation due to quantum fluctuations. A well known example is the electron–phonon coupling leading to an increased effective mass. In principle any fluctuation mode coupling to the Fermi liquid quasiparticle excitations can lead to such an effective mass enhancement. In order to explain in the case of $Sr_3Ru_2O_7$ the above described discrepancies between the zero field band structure observed by ARPES and the magnitude of the increase in the Sommerfeld constant, the fluctuation enhancement would have to be magnetic field dependent. Grigera et al. [8] established that the material is close to a quantum critical endpoint in phase space. The critical fluctuations associated with such a point in phase space would naturally give rise to a magnetic field dependent renormalisation of the band structure as a function of magnetic field. Most importantly, the mass renormalisation is expected to peak at the critical field. The fact that only part of the Fermi surface is significantly affected can in principle be understood if the critical fluctuations have a finite lattice momentum $q \neq 0$ and couple only to part of the

Fermi surface. Indeed incommensurate spin fluctuations with finite **q** have been observed by Capogna et al. [28] in neutron studies.

Quantum criticality in itinerant systems is a theoretically challenging subject. The best current approach to this field is the theoretical framework developed by Hertz [29] and Millis [30]. Two studies have applied the theory to the case of $Sr_3Ru_2O_7$ [20, 31]. Both predict the qualitative increase in the specific heat as a function of magnetic field, however, the detailed functional form differs. The exact behaviour of the specific heat as a function of magnetic field depends on the universality class of the quantum critical end point. Both papers predict a power law divergence of the Sommerfeld coefficient with the exponents being $-1/3$ [20] and $-1/2$ [31]. However, none of them treats the problem under the constraint of two spin species and number conservation, making a direct comparison to experiment impossible. It can be concluded that a quantum critical scenario would be generically consistent with the experimental observation of the onset of a power law divergence as a function of magnetic field as well as a mass enhancement of the single particle density of states peaked at the critical field of 7.9 T.

Finally, it has to be mentioned that another possibility of the observed enhancement of the specific heat is a thermal population of fluctuation modes. The resulting specific heat contribution would have to have a linear dependence on temperature. This would be from a theoretical point of view very unusual insofar as fluctuations give in general a specific heat with a non-linear temperature dependence due to their bosonic character. I am not aware of examples to the contrary in the literature, but cannot absolutely exclude this possibility.

In summary, in this section I discussed possible scenarios explaining the observed symmetric peak in the entropy of $Sr_3Ru_2O_7$ as a function of magnetic field around the critical field of 7.9 T. It was shown that without postulating a specific density of states containing very unusual features it is not possible to explain the magnitude and functional form of the singular contribution to the entropy solely in terms of the zero field density of states measured by ARPES. I furthermore discussed that allowing for critical fluctuations associated with the metamagnetic features one can naturally explain the existence of a power law as well as the significant enhancement of the electronic density of states as a function of distance from the critical field. More importantly, if these fluctuations are found not to significantly alter the electronic structure of $Sr_3Ru_2O_7$ then this places potentially strong constraints on theories regarding itinerant quantum critical end points.

References

1. Borzi RA, Grigera SA, Farrell J, Perry RS, Lister SJS, Lee SL, Tennant DA, Maeno Y, Mackenzie AP (2007) Formation of a nematic fluid at high fields in $Sr_3Ru_2O_7$. Science 315(5809):214–217
2. Berridge AM, Green AG, Grigera SA, Simons BD (2009) Inhomogeneous magnetic phases: a LOFF-like phase in $Sr_3Ru_2O_7$. Phys Rev Lett 102(13):136404

3. Grigera SA, Gegenwart P, Borzi RA, Weickert F, Schofield AJ, Perry RS, Tayama T, Sakakibara T, Maeno Y, Green AG, Mackenzie AP (2004) Disorder-sensitive phase formation linked to metamagnetic quantum criticality. Science 306(5699):1154–1157
4. Gegenwart P, Weickert F, Garst M, Perry RS, Maeno Y (2006) Metamagnetic quantum criticality in $Sr_3Ru_2O_7$ studied by thermal expansion. Phys Rev Lett 96(13):136402
5. Perry RS, Galvin LM, Grigera SA, Capogna L, Schofield AJ, Mackenzie AP, Chiao M, Julian SR, Ikeda SI, Nakatsuji S, Maeno Y, Pfleiderer C (2001) Metamagnetism and critical fluctuations in high quality single crystals of the bilayer ruthenate $Sr_3Ru_2O_7$. Phys Rev Lett 86(12):2661–2664
6. Binz B, Braun HB, Rice TM, Sigrist M (2006) Magnetic domain formation in itinerant metamagnets. Phys Rev Lett 96(19):196406
7. Condon JH (1966) Nonlinear de Haas–van Alphen effect and magnetic domains in beryllium. Phys Rev 145(2):526–535
8. Grigera SA, Perry RS, Schofield AJ, Chiao M, Julian SR, Lonzarich GG, Ikeda SI, Maeno Y, Millis AJ, Mackenzie AP (2001) Magnetic field-tuned quantum criticality in the metallic ruthenate $Sr_3Ru_2O_7$. Science 294(5541):329–332
9. Grigera SA, Borzi RA, Mackenzie AP, Julian SR, Perry RS, Maeno Y (2003) Angular dependence of the magnetic susceptibility in the itinerant metamagnet $Sr_3Ru_2O_7$. Phys Rev B 67(21):214427
10. Grigera SA. Private communication
11. Stryjewski E, Giordano N (1977) Metamagnetism. Adv Phys 26:487
12. Mercure J-F (2008) The de Haas–van Alphen effect near a quantum critical end point in $Sr_3Ru_2O_7$. PhD Thesis, University of St. Andrews
13. Damascelli A, Hussain Z, Shen ZX (2003) Angle-resolved photoemission studies of the cuprate superconductors. Rev Mod Phys 75(2):473–541
14. von Löhneysen H, Rosch A, Vojta M, Wölfle P (2007) Fermi-liquid instabilities at magnetic quantum phase transitions. Rev Mod Phys 79(3):1015–1075
15. Chubukov AV, Maslov DL, Gangadharaiah S, Glazman LI (2005) Thermodynamics of a Fermi liquid beyond the low-energy limit. Phys Rev Lett 95(2):026402
16. Ikeda S, Maeno Y, Nakatsuji S, Kosaka M, Uwatoko Y (2000) Ground state in $Sr_3Ru_2O_7$: Fermi liquid close to a ferromagnetic instability. Phys Rev B 62(10):R6089–R6092
17. Yamase H, Katanin AA (2008) Theory of spontaneous Fermi surface symmetry breaking for $Sr_3Ru_2O_7$. International Conference on Strongly Correlated Electron Systems (SCES 2007), Houston, TX, MAY 13–18, 2007. Phys B Condens Matter 403(5–9):1262–1264
18. Kee HY, Kim YB (2005) Itinerant metamagnetism induced by electronic nematic order. Phys Rev B 71(18):184402
19. Raghu S, Paramekanti A, Kim EA, Borzi RA, Grigera S, Mackenzie AP, Kivelson SA (2009) Microscopic theory of the nematic phase in $Sr_3Ru_2O_7$. Phys Rev B 79:214402
20. Millis AJ, Schofield AJ, Lonzarich GG, Grigera SA (2002) Metamagnetic quantum criticality in metals. Phys Rev Lett 88(21):217204
21. Ohmichi E, Yoshida Y, Ikeda SI, Mushunikov NV, Goto T, Osada T (2003) Double metamagnetic transition in the bilayer ruthenate $Sr_3Ru_2O_7$. Phys Rev B 67(2):024432
22. Perry RS, Kitagawa K, Grigera SA, Borzi RA, Mackenzie AP, Ishida K, Maeno Y (2004) Multiple first-order metamagnetic transitions and quantum oscillations in ultrapure $Sr_3Ru_2O_7$. Phys Rev Lett 92(16):166602
23. Borzi RA, Grigera SA, Perry RS, Kikugawa N, Kitagawa K, Maeno Y, Mackenzie AP (2004) De Haas–van Alphen effect across the metamagnetic transition in $Sr_3Ru_2O_7$. Phys Rev Lett 92(21):216403
24. Tamai A, Allan MP, Mercure JF, Meevasana W, Dunkel R, Lu DH, Perry RS, Mackenzie AP, Singh DJ, Shen Z-X, Baumberger F (2008) Fermi surface and van Hove singularities in the itinerant metamagnet $Sr_3Ru_2O_7$. Phys Rev Lett 101(2):026407
25. Iwaya K, Satow S, Hanaguri T, Shannon N, Yoshida Y, Ikeda SI, He JP, Kaneko Y, Tokura Y, Yamada T, Takagi H (2007) Local tunneling spectroscopy across a metamagnetic critical point in the bilayer ruthenate $Sr_3Ru_2O_7$. Phys Rev Lett 99(5):057208

References

26. Kikugawa N, Mackenzie AP, Bergemann C, Borzi RA, Grigera SA, Maeno Y (2004) Rigid-band shift of the Fermi level in the strongly correlated metal: $Sr_3Ru_2O_7$. Phys Rev B 70(6):060508
27. Farrell J, Perry RS, Rost A, Mercure JF, Kikugawa N, Grigera SA, Mackenzie AP (2008) Effect of electron doping the metamagnet $Sr_3Ru_2O_7$. Phys Rev B 78(18):180409
28. Capogna L, Forgan EM, Hayden SM, Wildes A, Duffy JA, Mackenzie AP, Perry RS, Ikeda S, Maeno Y, Brown SP (2003) Observation of two-dimensional spin fluctuations in the bilayer ruthenate $Sr_3Ru_2O_7$ by inelastic neutron scattering. Phys Rev B 67(1):012504
29. Hertz JA (1976) Quantum critical phenomena. Phys Rev B 14(3):1165–1184
30. Millis AJ (1993) Effect of a nonzero temperature on quantum critical-points in itinerant Fermion systems. Phys Rev B 48(10):7183–7196
31. Zhu LJ, Garst M, Rosch A, Si QM (2003) Universally diverging Grüneisen parameter and the magnetocaloric effect close to quantum critical points. Phys Rev Lett 91(6):066404

Chapter 6
Conclusions and Future Work

In this thesis a detailed study of the entropy of the itinerant metamagnet $Sr_3Ru_2O_7$ was presented. With the experimental setup specially developed for this work it was possible to study isothermal changes of the entropy as a function of magnetic field with the magnetocaloric effect as well as performing specific heat measurements. In the following, the main results of this project in relation to the physics of $Sr_3Ru_2O_7$ will be summarised. Particular attention will be paid to data relating to the novel quantum phase as well as the proposed quantum critical point and its effect on the surrounding normal states. The chapter will conclude with a brief discussion of planned future work and the wider relevance of the results.

$Sr_3Ru_2O_7$ has been intensely studied in recent years because of a proposed quantum critical point in its magnetic phase diagram [1] and the discovery of a novel quantum phase [3] which shows 'nematic-like' transport properties in its vicinity [2]. The first part of the results presented in this thesis particularly concerned the entropic properties of this new phase and the surrounding phase transitions. For clarity of the discussion I show again in Fig. 6.1 the magnetic field and temperature phase diagram for the magnetic field applied along the crystallographic c-axis. Its details are discussed in Sect. 2.2.2 (Chap. 2). Here it is only important that the data points are marking the position of proposed first (green) and second (blue) order phase transitions and crossovers as a function of magnetic field. In Chap. 5 it was shown that the entropy change and the jump in magnetisation across the first order transition (1) together with the curvature of the transition line in the H–T diagram are quantitatively consistent with the magnetic version of the Clausius–Clapeyron relation. This indicates that it is indeed a first order transition between two equilibrium phases. The same holds for the evolution of the entropy across transition (2) qualitatively. It was furthermore shown that the entropy of the anomalous phase region is higher than that of the surrounding low and high field states, indicating the existence of additional degrees of freedom.

Fig. 6.1 Phase boundaries of $Sr_3Ru_2O_7$ in the H–T plane with the magnetic field applied along the crystallographic c-axis. The data shown are extracted from magnetisation M, the complex AC magnetic susceptibility χ, linear magnetostriction λ and thermal expansion α [3, 4]. The *green lines* indicate first order transitions [(*1*) and (*2*)] and the *blue line* represents a second order phase transition. The other data represent crossovers

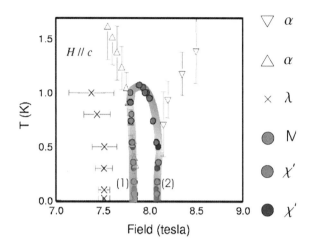

Furthermore, measurements of the specific heat as a function of temperature across the 'roof' feature, indicated by a blue line in Fig. 6.1, were performed. These results showed that it is consistent with a second order symmetry breaking phase transition at least for the case of the magnetic field being aligned parallel to the crystallographic c-axis. It was shown in previous experiments [5], that the high temperature state above the 'roof' is unusual in that its specific heat c divided by temperature T is consistent with having a logarithmically diverging contribution with a singularity at zero temperature at a critical field in the vicinity of 8 T. The experiments presented here showed that the low temperature novel quantum phase also has a highly unusual specific heat with c/T increasing linearly towards lower temperature and therefore being expressible as $c/T = \gamma - bT$. This is in particular significant since Mercure and co-workers [6] have established the existence of quantum oscillations inside the new phase indicating the existence of a Fermi surface and Landau Fermi liquid quasiparticles. However, the linear contribution to c/T to lowest temperatures seems inconsistent with standard Fermi liquid theory. It is therefore proposed here that either (a) finite temperature corrections to Fermi liquid theory [7] are important in this novel phase or that (b) additional non-Fermi liquid contributions to the specific heat exist. On the one hand, the detailed knowledge of the thermodynamic and microscopic properties of $Sr_3Ru_2O_7$ should enable a quantitative comparison of theoretical models of Fermi liquid corrections with experimental data. On the other hand, if the additional degrees of freedom of the second proposal are for example magnetic in nature, they are potentially measurable in neutron experiments.

Two important future experiments directly related to the ones presented here should be carried out to further study the points raised. First the measurements of specific heat in the novel quantum phase should be extended to below the minimum temperature of 200 mK. Though the experimental setup developed in this thesis can perform measurements in this temperature regime a reliable data analysis is constrained by the existence of a Schottky-type anomaly in the experimental addenda. This can potentially be removed in the next generation of the design. Secondly one should further extend the specific heat study to different angular orientations of the magnetic field. In Appendix B an initial angular study is presented that was carried out by the author. Here a specially cut sample with surfaces along different crystallographic orientations allowed the use of the same setup without the need of a rotation mechanism. This revealed a complex phase diagram for the magnetic field being applied at different angles within the *ab*-plane.

A detailed angular study is also required in order to study the angular dependence of the specific heat signature corresponding to the 'roof'. One way of achieving this is by adopting the setup for a novel three-axis vector magnet system currently being developed for the University of St Andrews. This will allow a smooth rotation of the magnetic field without friction heating as observed in more conventional rotation mechanisms. Here it should in particular be possible to address the question of whether an in-plane field changes the nature of the 'roof' from being a second order phase transition to a thermodynamic crossover.

The second part of the work presented here concerned the entropy of the low and high field states surrounding the anomalous phase. In Sect. 5.2 it is shown that both the specific heat and entropy increase significantly upon approaching the critical region in the phase diagram from either low or high magnetic field as has been inferred from previous measurements [1, 4]. However, at constant magnetic field, c/T is consistent with being a constant as a function of temperature over the temperature range explored. The data therefore indicate that the low and high field phases are Fermi liquids whose Sommerfeld coefficient sharply peaks at 7.9 T. It is scientifically important to address the issue of whether this peak can be understood solely in terms of a shift of the spin-up/spin-down Fermi energy through an otherwise rigid single particle density of states or if the single particle excitation spectrum is dominated by quantum fluctuations renormalising the quasiparticle properties as function of distance from the critical field at 7.9 T.

One of the main contributions of this thesis in relation to this issue is the discovery of a small Fermi surface pocket in quantum oscillations of the magnetocaloric effect contributing approximately 45% to the zero field specific heat. Inconclusive evidence for the existence of this pocket at the Fermi surface had also been seen in ARPES [8] together with the discovery that it is associated with a density of states peak originating in a saddle point van-Hove singularity energetically close to the Fermi energy. In particular the fact that no other Fermi surface sheet was observed in quantum oscillation measurements by Mercure and co-workers [6] to have a significant increase in effective mass and therefore specific heat contribution makes this new Fermi surface pocket the best candidate for being the primary cause of the metamagnetism and of the peak in entropy at the critical field.

The highly detailed measurements allow to show that the magnitude of the specific heat increase is at least 100% larger than estimated from the zero field density of states as measured by ARPES. Furthermore, from a qualitative point of view, the onset of divergence is consistent with a singular contribution to the entropy in the form of a scale invariant power law diverging at the critical field of 7.9 T. Such a power law does not naturally arise from a single feature in the band structure.

The results of this thesis therefore indicate that the peak in the entropy is significantly influenced either by a renormalisation of the quasiparticle properties due to quantum fluctuations or additional non-Fermi liquid degrees of freedom contributing to the specific heat. Both of these possibilities have their origin potentially in the underlying quantum critical point. Especially the first scenario of a peak in the density of states that is enhanced as a function of field would naturally produce a singularity in the entropy as a function of magnetic field having the form of a power law.

An important question is whether one can experimentally distinguish between the different scenarios. One possible experiment that has the potential to do so is the probing of the local density of states by scanning tunneling microscopy (STM) in magnetic fields. If indeed there is a peak in the electronic density of states of $Sr_3Ru_2O_7$ that is fluctuation enhanced as a function of distance from the critical field, then both the existence of the peak and its field dependent change in magnitude should in principle be observable in the current–voltage characteristic of a STM measurement. In the case of no fluctuation enhancement one should at least still observe a peak of the current–voltage characteristic near zero bias corresponding to the proposed van Hove singularity.

The work presented in this thesis should be considered in its wider context. The detailed knowledge of the thermodynamic properties of $Sr_3Ru_2O_7$ in combination with its suitability for nearly all major microscopic solid state probes make the material an exceptional example in which the effect of electron correlations can be studied. In particular it can serve as an important quantitative test case for theories relating to the physics of itinerant quantum critical end points and 'electron-nematic' phases.

Furthermore, from an experimentalist's point of view the method of extracting the isothermal changes in the entropy of a system as a function of magnetic field with the magnetocaloric effect in the non-adiabatic limit has been very successful. First of all, it allowed the careful study of entropy changes across magnetic first order transitions and to quantitatively test the Clausius–Clapeyron relation. Secondly, the technique allowed for the quantitative measurement of entropy changes in a sample as a function of magnetic field without the extrapolation of specific heat measurements to zero temperature and the uncertainties involved with this. This experimental method should easily be applicable to a range of other materials whose magnetic phase diagrams indicate the presence of a quantum critical point or have novel quantum states such as $YbRh_2Si_2$ [9] or URhGe [10]. In doing so it will be possible to study a range of different classes of quantum critical points for which the experimentally measured evolution of entropy with magnetic field can be tested against theory.

Finally it should be mentioned that the observation of quantum oscillations in the magnetocaloric effect has been shown to not only reproduce the results of other quantum oscillation measurements but also be particularly suited to the investigation of low frequency signals and the detection of small Fermi surface pockets with frequencies below 1 kT. It should therefore be extended to materials where such small Fermi surface areas are predicted by band structure calculations or otherwise suspected to be crucial to the material properties, such as the recently observed small Fermi surface pockets in the cuprate high temperature superconductors [11–13].

References

1. Grigera SA, Perry RS, Schofield AJ, Chiao M, Julian SR, Lonzarich GG, Ikeda SI, Maeno Y, Millis AJ, Mackenzie AP (2001) Magnetic field-tuned quantum criticality in the metallic ruthenate $Sr_3Ru_2O_7$. Science 294(5541):329–332
2. Borzi RA, Grigera SA, Farrell J, Perry RS, Lister SJS, Lee SL, Tennant DA, Maeno Y, Mackenzie AP (2007) Formation of a nematic fluid at high fields in $Sr_3Ru_2O_7$. Science 315(5809):214–217
3. Grigera SA, Gegenwart P, Borzi RA, Weickert F, Schofield AJ, Perry RS, Tayama T, Sakakibara T, Maeno Y, Green AG, Mackenzie AP (2004) Disorder-sensitive phase formation linked to metamagnetic quantum criticality. Science 306(5699):1154–1157
4. Gegenwart P, Weickert F, Garst M, Perry RS, Maeno Y (2006) Metamagnetic quantum criticality in $Sr_3Ru_2O_7$ studied by thermal expansion. Phys Rev Lett 96(13):136402
5. Perry RS, Galvin LM, Grigera SA, Capogna L, Schofield AJ, Mackenzie APChiao M, Julian SR, Ikeda SI, Nakatsuji S, Maeno Y, Pfleiderer C (2001) Metamagnetism and critical fluctuations in high quality single crystals of the bilayer ruthenate $Sr_3Ru_2O_7$. Phys Rev Lett 86(12):2661–2664
6. Mercure J-F (2008) The de Haas–van Alphen effect near a quantum critical end point in $Sr_3Ru_2O_7$. Ph.D. thesis, September 2008.
7. Condon JH (1966) Nonlinear de Haas–van Alphen effect and magnetic domains in beryllium. Phys Rev 145(2):526–535
8. Tamai A, Allan MP, Mercure JF, Meevasana W, Dunkel R, Lu DH, Perry RS, Mackenzie AP, Singh DJ, Shen Z-X, Baumberger F (2008) Fermi surface and van Hove singularities in the itinerant metamagnet $Sr_3Ru_2O_7$. Phys Rev Lett 101(2):026407
9. Custers J, Gegenwart P, Wilhelm H, Neumaier K, Tokiwa Y, Trovarelli O, Geibel C, Steglich F, Pepin C, Coleman P (2003) The break-up of heavy electrons at a quantum critical point. Nature 424(6948):524–527
10. Lévy F, Sheikin I, Grenier B, Huxley AD (2005) Magnetic field-induced superconductivity in the ferromagnet UR1hGe. Science 309(5739):1343–1346
11. Doiron-Leyraud N, Proust C, LeBoeuf D, Levallois J, Bonnemaison J-B, Liang R, Bonn DA, Hardy WN, Taillefer L (2007) Quantum oscillations and the Fermi surface in an underdoped high-T_c superconductor. Nature 447(7144):565–568

12. Sebastian SE, Harrison N, Palm E, Murphy TP, Mielke CH, Liang R, Bonn DA, Hardy WN, Lonzarich GG (2008) A multi-component Fermi surface in the vortex state of an underdoped high-T_c superconductor. Nature 454(7201):200–203
13. Yelland EA, Singleton J, Mielke CH, Harrison N, Balakirev FF, Dabrowski B, Cooper JR (2008) Quantum oscillations in the underdoped cuprate $YBa_2Cu_4O_8$. Phys Rev Lett 100(4):047003

Chapter 7
Appendices

7.1 Appendix A: Material Properties

The form and dimensions of the materials in the experimental setup are detailed in Table 7.1. Silver is used for the sample platform. The Kevlar strings are supporting the sample platform with electrical contacts to the thermometer and heater made via the Manganin wires. The platinum wire is used as a tuneable thermal link between the sample and the thermal bath. Furthermore listed are the approximate dimensions of the sample used as well as the various amorphous materials necessary for positioning thermometer, heater and sample.

Relevant literature values for the low temperature specific heat and thermal conductivity of materials used in the experimental setup are given in Table 7.2. The thermal conductivities of silver and platinum depend significantly on the impurity and defect concentration of the material and can differ from sample to sample.

7.2 Appendix B: Angular Dependence of the Magnetocaloric Signal

In Chap. 5, I presented the main results regarding the entropic properties of $Sr_3Ru_2O_7$ as a function of magnetic field and temperature with the field applied parallel to the crystallographic c-axis. However, as previously discussed, $Sr_3Ru_2O_7$ also has a very interesting phase diagram for other field orientations. Here I will give an overview of the most significant results as obtained from a preliminary angular study. In order to achieve this with the current setup the specially cut sample C698K was used.

Figure 7.1a shows the sample dimensions. The surface marked with (A) is parallel to the crystallographic ab plane and has an octagonal cross section. The

Table 7.1 Form and dimensions of the materials in the experimental setup

Material	Length l (mm)	Cross section A
Silver (cuboid)	0.15	4 mm × 4 mm = 16 mm^2
Kevlar (eight strings of 35 filaments)	16	$8 \times 35 \times \pi \times 0.017$ mm^2 = 0.25 mm^2
Manganin (four twisted pairs)	20	$4 \times 2 \times \pi \times 0.030$ mm^2 = 0.023 mm^2
Platinum (one wire)	20	$\pi \times 0.025$ mm^2 = 0.0020 mm^2
Sr$_3$Ru$_2$O$_7$ (cuboid)	1.5	1.5 mm × 1.5 mm = 2.25 mm^2
GE 7031		overall < 1 mg
Apiezon N		overall < 1 mg
Stycast		overall < 1 mg

Table 7.2 Relevant literature values for the low temperature specific heat and thermal conductivity of materials used in the experimental setup

Material	Specific heat c	Thermal conductivity κ
Silver	7×10^{-4} J/mol K [1]	25 W/m K [2]
Kevlar	–	αT^n [3] $\alpha = 3.9 \times 10^{-5}$ W/cm K^{1+n} $n = 1.71$
Manganin	$sT^{-2} + aT + bT^3 + cT^5$ [4] $s = 36$ μJ K^2/g $a = 56$ μJ/g K^2 $b = 6.02$ μJ/g K^4 $c = -0.029$ μJ/g K^6	αT^n [5] $\alpha = 9.5 \times 10^{-4}$ W/cm K^{1+n} $n = 1.19$
Platinum	–	3 W/m K [2, 6]
Stycast	$aT + bT^3 + cT^5$ [7] $a = 2.91$ μJ/g K^2 $b = 15.7$ μJ/g K^4 $c = 8.98$ μJ/g K^6	$aT^{1.98}$ [8] $a = 4.9 \times 10^{-4}$ W/cm K^2
Apiezon N	$aT + bT^3$ [9] $a = 1.32$ μJ/g K^2 $b = 25.8$ μJ/g K^4	–

surface opposite to (A) is cut such that its normal forms an angle of 19° with the crystallographic c axis. Figure 7.1b shows a reflection Laue pattern taken of surface (A). The borders of the surface are not parallel to the crystallographic axes. Nevertheless one is closer to the a or b^1 direction (these are equivalent in Sr$_3$Ru$_2$O$_7$) and one is closer to the 45° direction. This sample can now be mounted in the three additional ways on the setup as shown in Fig. 7.1c allowing magnetocaloric sweeps at an angle of 19° between the magnetic field and the c-axis as

---

[1] The crystallographic axes here are not those of the orthorhombic structure but of the idealised tetragonal unit cell. In particular they are along the direction of the nearest neighbor Ru–O–Ru bond and therefore at 45° to the true crystal axes of the orthorhombic unit cell. This notation is chosen in order to be consistent with the paper by Borzi et al. [10].

7.2 Appendix B: Angular Dependence of the Magnetocaloric Signal

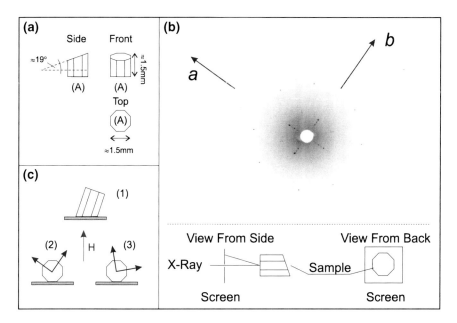

Fig. 7.1 Overview of sample C698K, which has been specially prepared for an initial angular dependence study of the magnetocaloric signal. **a** The front, side and top view of the sample. The base is a regular octagon that is parallel to the crystallographic *ab* plane. The top of the sample has been cut such that the normal to the resulting surface is forming an angle of 19° with the crystallographic *c*-axis. The overall dimensions are as given in the figure. **b** A reflection Laue picture of the sample. The geometry of the X-ray beam, screen and sample are given below the photograph. One can clearly identify the crystallographic *a* and *b* axis. The sides of the octagon are rotated by approximately 10° with respect to the crystal axes. **c** The three different ways in which the sample can be mounted on the sample platform (*grey*). The *blue arrows* are indicating the crystallographic axes and the direction of the magnetic field *H* is shown in *red*. In orientation (*1*) the applied magnetic field is at 19° to the *c*-axis. In orientation (*2*) on the other hand the field is in the *ab* plane and 35° off the *a/b* axis. Finally in orientation (*3*) the magnetic field is again in the *ab* plane but this time 10° off the *a/b* axis

well as two measurements with the magnetic field perpendicular to the *c*-axis but at different angles in the *ab*-plane.

7.2.1 Study at 19°

Figure 7.2 shows the temperature of the sample as a function of magnetic field during a field sweep at 0.02 T/min (blue) and −0.02 T/min (red) at a sample temperature of 160 mK. The inset on the upper right is showing the orientation of the sample. Coming from low fields the metamagnetic crossover is still clearly identifiable but has moved to 6.9 T compared to 7.5 T in the case of the magnetic field being applied parallel to the *c*-axis.

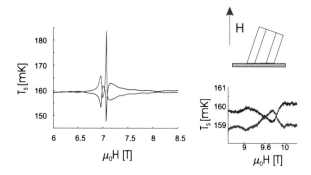

Fig. 7.2 The graph is showing the temperature T_S of the sample during a sweep of the magnetic field H at 0.02 T/min (*blue curve*) and −0.02 T/min (*red curve*). The orientation of the sample during the sweep is shown in the *upper right inset*. The *lower right inset* is showing the same property as the main graph but in the field range of the high field crossover

Similarly the main first order transition has changed from 7.9 T to 7.1 T. The second first order phase transition is no longer identifiable.

The second inset on the lower right shows the position of the high field magnetocaloric crossover that has moved from 12.5 T to 9.5 T.

7.2.2 Study in the ab-Plane

Two experiments were conducted with the magnetic field being applied in the *ab* plane. The first shown in the upper right part of Fig. 7.3a is such that the magnetic field is oriented in the same plane as is defined by the crystallographic *c*-axis and the orientation of the '19°' experiment reported above. The second orientation shown in the upper left part of Fig. 7.3b is achieved by rotating the crystal by 45° around its *c*-axis. It is this second orientation where the applied magnetic field points closer along a crystallographic *a* or *b* axis.[2]

In Fig. 7.3a the temperature of the sample during an up (blue) and down (red) sweep of the magnetic field for orientation (2) is shown. The analogous Fig. 7.3b is showing the equivalent sweeps for the orientation (3). The difference in the measurements reveal a significant anisotropy of the phase diagram for field orientations in the *ab*-plane.

First of all, although both show signatures of the expected metamagnetic first order transition (I), this is situated at different fields, namely 5.3 T for orientation (2) and 5.1 T for orientation (3). Furthermore while the magnetocaloric

---

[2] As discussed earlier these are not distinguishable by the experiments reported here, which is why the labelling is to some extend arbitrary.

7.2 Appendix B: Angular Dependence of the Magnetocaloric Signal

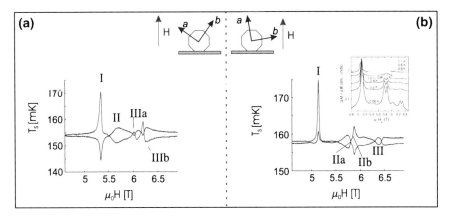

Fig. 7.3 In both figure **a** and **b** the temperature of the sample T_S during a sweep of the magnetic field at 0.02 T/min (blue) and −0.02 T/min (red) is shown as a function of magnetic field H. The schematics in the centre are giving the orientation of the magnetic field (red) relative to the crystallographic a/b-axis (blue). Furthermore the *inset* of figure **b** is showing a previous measurement of dM/dB as a function of magnetic field for a similar field orientation as the presented magnetocaloric sweep [11]

signal for the up and down sweep in orientation (2) is approximately symmetric as expected for a magnetocaloric signal, it is highly asymmetric in case (3). This implies that in orientation (3) there always occurs heating of the sample independent of the direction in magnetic field in which the transition is crossed. One explanation of this effect would be that due to magnetostriction in the material friction heating is caused either by dislocations in the sample or at the interface between the sample and the sample holder. However, no ageing effect could be observed after repeating the experiment at the same temperature for several times. Neither does the relative size of the signature between up and down sweep change significantly by changing the sweep rate by a factor of 4. The fact however that there is a strong asymmetry between the up and down sweep indicates that there is nevertheless a significant latent heat contribution associated with the transition.

Secondly between 5.5 T and 6.5 T several peaks in the sample as a function of magnetic field indicate a series of phase transitions or crossovers. Though a definite identification without further angular studies is not possible, the most likely is the one indicated in the two graphs. In this scenario, feature (II) in Fig. 7.3b is upon changing the orientation of the magnetic field split into features (IIa) and (IIb) in Fig. 7.3c. Features (IIIa) and (IIIb) in Fig. 7.3b on the other hand are nearly merged into one feature in Fig. 7.3c. Finally, the inset in Fig. 7.3c shows a susceptibility measurement published by Perry et al. for the magnetic field applied along the crystallographic a direction. This is close to orientation (2) reported here and as can be seen the position of features in the current experiment are well reflected in the susceptibility measurement.

7.2.2.1 Discussion

These measurements represent a clear indication that the magnetic phase diagram of $Sr_3Ru_2O_7$ has a strong dependence on the orientation of the magnetic field direction in the crystallographic *ab*-plane. Whereas experiments cannot distinguish between the magnetic field being applied along the *a* (*[100]*-direction) or *b* (*[010]*-direction) directions it is clear that the physics for the field applied along the *[110]*-direction is systematically different. In [12], Mercure and co-workers determined that the results for the angular dependence of the de Haas-van Alphen measurements with the magnetic field close to the *c*-axis (*[001]*) can best be described by assuming that the strength of the coupling between the spin and the magnetic field is anisotropic at high fields. Mathematically this can be done by assuming that the usually scalar Landé *g*-factor, describing the strength of coupling between the spin **S** and the magnetic field **H** via the term $\hat{H}_{mag} = -g\mathbf{SH}$, is replaced by a positive definite 3×3 matrix **G**, leading to a revised contribution to the Hamiltonian of the form $\hat{H}_{mag} = -\mathbf{S}(\mathbf{GH})$. One has to remember that experiments at low fields showed an isotropic Pauli paramagnetism [13] and that the above description can only be valid in high fields. The experiments here however suggest, that the magnetic field positions of transitions in the zero temperature limit are not simply scaled by an anisotropic **G** factor, but rather that the phase diagram is affected in a fundamentally nonlinear way.

There are several possible reasons for this behaviour. First of all it is known that there is a strong magnetoelastic coupling present in the material. Therefore, applying the magnetic field along different symmetry axes of the crystalline structure can in principle lead to small but different structural distortions. Since this affects the electronic structure of the material, it is a possible source for the observed anisotropy. This would also explain why the anisotropy is absent at zero field, since the energetic effects of such Fermi surface distortions are undoubtedly of higher order in magnetic field. A second possible reason for the observed anisotropy is spin orbit coupling. This effect has in particular been discussed for the two single layer compounds Sr_2RhO_4 [14, 15] and Sr_2IrO_4 [16], which are isostructural to Sr_2RuO_4.

However a more detailed angular study for the magnetic field being aligned in the *ab*-plane is necessary to study this complex and rich phase diagram in detail. For this purpose a vector magnet system is currently being put in place and an advanced version of the setup presented in this thesis is being developed.

References

1. du Chatenier FJ, Miedema AR (1965) Heat capacity below 1 K: observation of the linear term and the HFS contribution in some dilute alloys. Proceedings of the 9th International Conference on Low Temperature Physics, p 1029
2. Pobell F (1992) Matter and methods at low temperatures. Springer-Verlag,Heidelberg

3. Ventura G, Barucci M, Gottardi E, Peroni I (2000) Low temperature thermal conductivity of Kevlar. Cryogenics 40(7):489–491
4. Marklund K, Bystrom S, Larsson M, Lindqvis T (1973) Low-temperature heat-capacity of Manganin wire and Woods metal. Cryogenics 13(11):671–672
5. Peroni I, Gottardi E, Peruzzi A, Ponti G, Ventura G (1999) Thermal conductivity of Manganin below 1 K. Nuclear Physics B—Proceedings Supplements 78:573–575. 6th International Conference on Advanced Technology and Particle Physics, Villa Olmo, Italy, October 05–09, 1998
6. Davey G, Mendelssohn K, Sharma JKN (1965) In: Proceedings of the Ninth International Conference on Low Temperature Physics, p 1196
7. Siqueira ML, Rapp RE (1991) Specific heat of an epoxi resin below 1 K. Rev Sci Instrum 62(10):2499–2500
8. Armstrong G, Greenberg AS, Sites JR (1978) Very low temperature thermal conductivity and optical properties of Stycast 1266 epoxy. Rev Sci Instrum 49(3):345–347
9. Schink HJ, Löhneysen Hv (1981) Specific-heat of Apiezon-N grease at very low-temperatures. Cryogenics 21(10):591–592
10. Borzi RA, Grigera SA, Farrell J, Perry RS, Lister SJS, Lee SL, Tennant DA, Maeno Y, Mackenzie AP (2007) Formation of a nematic fluid at high fields in $Sr_3Ru_2O_7$. Science 315(5809):214–217
11. Perry RS, Tayama T, Kitagawa K, Sakakibara T, Ishida K, Maeno Y (2005) Investigation into the itinerant metamagnetism of $Sr_3Ru_2O_7$ for the field parallel to the ruthenium oxygen planes. J. Phys. Soc. Jpn 74(4):1270–1274
12. Mercure J-F (2008) The de Haas-van Alphen effect near a quantum critical end point in $Sr_3Ru_2O_7$. PhD Thesis. September 2008
13. Ikeda S, Maeno Y, Nakatsuji S, Kosaka M, Uwatoko Y (2000) Ground state in $Sr_3Ru_2O_7$: Fermi liquid close to a ferromagnetic instability. Phys Rev B 62(10):R6089–6092
14. Haverkort MW, Elfimov IS, Tjeng LH, Sawatzky GA, Damascelli A (2008) Strong spin-orbit coupling effects on the fermi surface of Sr_2RuO_4 and Sr_2RhO_4. Phys Rev Lett 101(2):026406
15. Liu G-Q, Antonov VN, Jepsen O, Andersen OK (2008) Coulomb-enhanced spin-orbit splitting: the missing piece in the Sr_2RhO_4 puzzle. Phys Rev Lett 101(2):026408
16. Kim BJ, Jin H, Moon SJ, Kim J-Y, Park J-Y, Leem CS, Yu J, Noh TW, Kim C, Oh S-J, Park J-H, Durairaj V, Cao G, Rotenberg E (2008) Novel J_{eff}=1/2 Mott state induced by relativistic spin-orbit coupling in Sr_2IrO_4. Phys Rev Lett 101(7):076402